Andrea M. Bludowsky

Optimal Crossover Designs with Interactions between Treatments & Units

Andrea M. Bludowsky

Optimal Crossover Designs with Interactions between Treatments & Units

Results from maximizing the trace of the information matrix of a design.

Südwestdeutscher Verlag für Hochschulschriften

Impressum/Imprint (nur für Deutschland/ only for Germany)
Bibliografische Information der Deutschen Nationalbibliothek: Die Deutsche Nationalbibliothek verzeichnet diese Publikation in der Deutschen Nationalbibliografie; detaillierte bibliografische Daten sind im Internet über http://dnb.d-nb.de abrufbar.

Alle in diesem Buch genannten Marken und Produktnamen unterliegen warenzeichen-, marken- oder patentrechtlichem Schutz bzw. sind Warenzeichen oder eingetragene Warenzeichen der jeweiligen Inhaber. Die Wiedergabe von Marken, Produktnamen, Gebrauchsnamen, Handelsnamen, Warenbezeichnungen u.s.w. in diesem Werk berechtigt auch ohne besondere Kennzeichnung nicht zu der Annahme, dass solche Namen im Sinne der Warenzeichen- und Markenschutzgesetzgebung als frei zu betrachten wären und daher von jedermann benutzt werden dürften.

Verlag: Südwestdeutscher Verlag für Hochschulschriften Aktiengesellschaft & Co. KG
Dudweiler Landstr. 99, 66123 Saarbrücken, Deutschland
Telefon +49 681 37 20 271-1, Telefax +49 681 37 20 271-0
Email: info@svh-verlag.de
Zugl.: Dortmund, TU Dortmund, Diss., 2009

Herstellung in Deutschland:
Schaltungsdienst Lange o.H.G., Berlin
Books on Demand GmbH, Norderstedt
Reha GmbH, Saarbrücken
Amazon Distribution GmbH, Leipzig
ISBN: 978-3-8381-1299-2

Imprint (only for USA, GB)
Bibliographic information published by the Deutsche Nationalbibliothek: The Deutsche Nationalbibliothek lists this publication in the Deutsche Nationalbibliografie; detailed bibliographic data are available in the Internet at http://dnb.d-nb.de.

Any brand names and product names mentioned in this book are subject to trademark, brand or patent protection and are trademarks or registered trademarks of their respective holders. The use of brand names, product names, common names, trade names, product descriptions etc. even without a particular marking in this works is in no way to be construed to mean that such names may be regarded as unrestricted in respect of trademark and brand protection legislation and could thus be used by anyone.

Publisher: Südwestdeutscher Verlag für Hochschulschriften Aktiengesellschaft & Co. KG
Dudweiler Landstr. 99, 66123 Saarbrücken, Germany
Phone +49 681 37 20 271-1, Fax +49 681 37 20 271-0
Email: info@svh-verlag.de

Printed in the U.S.A.
Printed in the U.K. by (see last page)
ISBN: 978-3-8381-1299-2

Copyright © 2010 by the author and Südwestdeutscher Verlag für Hochschulschriften Aktiengesellschaft & Co. KG and licensors
All rights reserved. Saarbrücken 2010

Contents

List of Figures   iii

List of Tables   v

Acknowledgements   vii

1 Introduction   1

2 The Crossover Design and an Instrument for Finding Optimal Designs   5
   2.1 The Model and its Information Matrix C_d 5
   2.2 A Method for Finding Optimal Designs   7
      2.2.1 The Universal Optimum   7
      2.2.2 Maximizing the Trace of C_d 8

3 Deriving General Formulas and Some of Their Properties   13
   3.1 Definitions of Variables   13
   3.2 Deriving $c_{11}(l), c_{12}(l)$ and $c_{22}(l)$ 15
   3.3 Some Properties of $c_{11}(l)$ and $c_{12}(l)$ 17
   3.4 Special Sequence Class Functions h_l 26

4 Optimal Designs   29
   4.1 Sequence Length $p = 3$ 29
   4.2 Sequence Length $p = 4$ 33
   4.3 Sequence Length $p = 5$ 40
   4.4 Sequence Length $p = 6$ 53

5 Conclusions and Recommendations   69

Appendices		73
A : Notations		73
B : Equivalence Classes of Treatments		77
B.1 Sequence Length $p = 5$		77
B.1.1 Set of all Equivalence Classes		77
B.1.2 Steps of Argumentation		80
B.2 Sequence Length $p = 6$		86
B.2.1 Set of all Equivalence Classes		86
B.2.2 Steps of Argumentation		96
C : Some more Lemmata and Transformations		121
C.1 Lemmata		121
C.2 Term Transformations		127
Bibliography		129

List of Figures

2.1 $q_d(x)$ is a linear combination of two h_l functions. 11
2.2 $q_d(x)$ only consists of h_1. The $\min_x \max_l h_l$ is at the minimum of the h_1 function. 11

4.1 Proportions $\alpha(\gamma)$ of equivalence class 2 sequences for an approximate optimal design with $p = 3$ periods. 33
4.2 Proportions $\alpha(\gamma)$ of equivalence class 2 sequences for an approximate optimal design with $p = 4$ periods. 40
4.3 Sequence proportions $\alpha(\gamma)$ and $\beta(\gamma)$ of equivalence classes 2 and 19, respectively, for an approximate optimal design with $p = 5$ periods. 53
4.4 Sequence proportions $\alpha(\gamma)$ of equivalence class 2 and $\beta(\gamma)$ of equivalence classes 19 and 28, respectively, for an approximate optimal design with $t = p = 5$ and $t = p = 6$ periods. 68

List of Tables

3.1 Matrix entries of S_{du}^{-1}, dependent on the number of treatment-j-appearances $n_j(l)$. 14

4.1 Equivalence classes l, their representative sequences [...] and h_l functions for sequence length $p = 3$. 30

4.2 Equivalence classes l, their representative sequences [...] and h_l functions for sequence length $p = 4$. 34

4.3 Some equivalence classes l, their representative sequences [...] and h_l functions for sequence length $p = 5$. 41

4.4 Some equivalence classes l, their representative sequences [...] and h_l functions for sequence length $p = 6$. 55

B.1 All equivalence classes, their representative sequences [...] and h_l functions for sequence length $p = 5$. 80

B.2 All equivalence classes, their representative sequences [...] and h_l functions for sequence length $p = 6$. 96

Acknowledgments

Even thought this thesis just bears my name, it was accomplished by a lot of support of many people. My gratitude is for all the people that contributed:

- First, thanks to Prof. Dr. J. Kunert. His support for a scholarship in the "Graduiertenkolleg Statistische Modellbildung" was the opportunity to earn a doctorate. Furthermore, he has given the inspiration for the topic of my thesis and assisted me with his academic advise.

- Thanks to Prof. Dr. C. Weihs. He has provided me with necessary assistance for clear and coherent proofs of my mathematical work.

- I would like to thank Prof. Dr. G. Trenkler for taking the time in reading the thesis twice and helping me to improve my writing skills.

- Thanks to my friends Vivien Rothe and Simone Wenzel for their motivating talks in times of throwbacks.

- Special thanks are also given to my beloved husband, Thomas Bludowsky, for the hours of taking a walk with our baby-daughter Emma. Thus, I had time to concentrate on my work or to recover from many sleepless nights.

- My last gratitude goes to my parents Marlies and Bernd Preußer, as their help and support for my educational career was the best I could think of.

Andrea Bludowsky

1 Introduction

In product development, e.g. food industry, it is important to adapt quality standards efficiently, taking into account the changing demands of consumers. For production lines, it is essential to perform tests in sensory profiling in order to control quality. Periodical tests of flavor are helpful to detect defective products of, e.g., the daily production. Further, they adapt the sensory product quality to the changing preferences of consumers. The application of statistical methods in sensory profiling attracts growing interest in theory and industrial practice: Lundahl & McDaniel (1988) discuss the panelist effect to be random rather than fixed, Næs (1991) explores the handling of individual differences between assessors, or Brockhoff (1998) considers a special model to compare panels.

Usually, sensory experiments are organized as follows: A set of assessors, called the panel, evaluates samples of products in several periods of time. In the language of design, the assessors are identified as blocks or units. The samples of different products are identified as treatments. The experiment uses a crossover design with t treatments being observed by n units, each having a length of p periods. Crossover designs are especially useful if there is an assumed or known variability among the units. The comparison of treatment effects can be done on the same unit.

Many papers deal with the most common model for crossover designs which include period, unit, direct treatment and carryover effects, e.g. Stufken (1996) or Afsarinejad & Hedayat (2002). However, as Han (2007) points out, some authors assess the common model to be too simple in its assumptions. Matthews (1988) suggests that unit×treatment interaction may be of additional importance. Therefore, special interest is paid to the interaction between products and assessors in this thesis. The interactions represent some kind of the assessors preferences for certain products, e.g. some may favor chocolate flavor. Thus, unit×treatment interaction is apparently existent in tests of preference. The carryover effects represent the influence of adjacent period treatments on the treatment evaluation of the current time period. Carryover effects are caused, e.g., through lingering taste so that a sweet apple juice seems to be bitter after a grape juice. Certainly, such a carryover effect affects subsequent food products. There

1 Introduction

are very few papers available dealing with the determination of optimal crossover designs including both effects: unit×treatment interaction and carryover effects. Thus, the focus of this thesis is set on this aspect.

The parametric model, being discussed in chapter 2.1, is, as already mentioned, motivated from analyzing sensory panel data. Analysis of Variance (ANOVA) is a technique that accounts for individual differences and is frequently used for analyzing sensory profile data. Testing hypotheses in an ANOVA is done by creating F-Ratios that are related to the analyzed subject. In detail, this relation compares a mean square of the interesting source of variation (between-treatment variability) with its expected value observed within the null hypothesis (within-treatments variability). However, there are differences in the denominator if the panelist effect is either considered to be random or fixed, i.e., the denominator would be the mean square of the interaction effect or the mean square of the error, respectively.

In contrast to authors like O'Mahony (1986), the problem of the denominator of the F-Ratio is usually solved by accepting assessors as random effects. This point of view is taken because the assessors will always represent some type of population, at least those people who would have passed the same training as the actual assessors in the panel. Authors as Næs & Langsrud (1998) often argue that conclusions from a fixed model (treatments are also considered to be fixed) are only valid for the particular set of individuals participating. In addition to random assessor effects, the model implies no constraints concerning the interaction effects among assessors and products.

The interest of this thesis is to obtain a structure of an optimal design for the given model. A descriptive example may be the evaluation of $t = 4$ different kinds of coffee, named 1, 2, 3 and 4. A variety of n probands is supposed to evaluate $p = 4$ samples of coffee each. The question of optimality is, which sequence is the best to serve the coffee samples to the probands in the four time periods. Suppose each proband evaluates each kind of coffee once. Thus, a proband would get a sequence of coffee samples equivalent to $[1, 2, 3, 4]$, i.e., there is a different sample of coffee for each time period. Another sequence may also be a sequence like $[1, 2, 3, 3]$, in which the last two coffee samples are identical. The advantage of direct repetition is the unbiased estimation of the carryover effect having an influence on the evaluation of the repeated treatment. There are plenty of possible sequences occurring in such an experiment. But most certainly it will not be manageable to perform all possibilities of such sequences, especially if the number of time periods p is growing in magnitude. The statistical design of experiments is a tool to determine a maximum of significant information with as few as possible experiments and to find the optimal parameters for a given process. An optimal design is the best out of a set of possible designs. Its optimality is measured by a certain criterion. There are many different such criteria of optimality. A design that is optimal under a multitude of criteria is called universally optimal. Kemmler (1990) presented an optimal design for the case of $p = 2$ periods in a model with carryover effects and random unit×treatment interaction. However, constructing efficient designs for larger p appears to be very demanding and time consuming for complex models with carryover effects, if it is possible at all. A huge improvement was provided by Kushner (1997). He introduced a method which simplifies the construction of an optimal design through maximization of the trace of the information matrix on the basis of quadratic, design-dependent functions. Kunert & Martin (2000a), e.g., managed to generalize this method for interference models. Moreover, optimal crossover designs have already been presented for a model with self and mixed carryover effects, see Kunert & Stufken (2002). Referring to the model with carryover and interaction effects, this thesis covers the cases of structure analysis for optimal designs with $p = 3$, 4, 5, and 6 periods, as those are the most practical sequence lengths used in experiments of sensory studies.

The method by Kushner (1997) is presented in section 2.2. It is used to examine whether interactions among assessors and products influence the optimality of common designs. Furthermore, Kushner's method is applied in order to evaluate how interaction and carryover effects operate conjointly. In a traditional model without assessor×product interaction, an optimal design is given by a combination of sequences that have no replication in product samples at all (their proportion of sequences in the design is $[(p-1)t - 1]/[(p-1)t]$), and sequences that repeat the same product in the last two samples of the experiment (with proportion $1/[(p-1)t]$ of sequences in the design). If the coefficient γ of the variance of the interaction effect increases, a decreasing proportion of sequences with one replicate in the last period and an increasing proportion of sequences with no replication of treatments would be conceivable. Of course, the coefficient γ must be known. To this end, the dissertation of Han (2007) proposes the restricted maximum likelihood estimation for the variance components in a model that includes unit×treatment interaction, among other things.

In order to obtain the desirable characteristics mentioned above, some formulas of auxiliary functions have to be derived. The auxiliary functions and their properties for certain parameter performances are discussed in chapter 3. Their applications for finding an optimal design in 3-, 4-, 5- and 6- time periods experiments are described in chapter 4. Conclusions and recommendations for possible research extending the results derived in this thesis are given in chapter 5. A list of notations as well as some supporting technical arguments and results are provided in the appendices.

1 Introduction

2 The Crossover Design with Interaction and Carryover Effects and an Instrument for Finding Optimal Designs

This chapter introduces the parametric model of a crossover design with interaction between products and assessors. Furthermore, the method by Kushner (1997) is presented and applied to the model. For a better understanding, all important mathematical abbreviations and notations are enlisted and described in Appendix A.

2.1 The Model and Its Information Matrix C_d

A design is the basis of every scientific experiment. A general block design prescribes how the treatments are to be assigned to the block plots. In sensory experiments, it is common to compare effects in human subjects (= units) that receive several treatments in different time periods. The attribute of one subject receiving different treatments is what crossover designs account for. Typically, the primary goal is to compare the treatment effects. The treatment effects are measured on the same subject in crossover designs. Thus, they are favored amongst others, especially if there is naturally variability among the subjects, see Cox (1958).

Let d be a design assigning treatment $d(u,r)$ to time period r of unit u. The set of all designs with t treatments, n units and $p \leq t$ periods is defined as $\Omega_{t,n,p}$. For $d \in \Omega_{t,n,p}$ the response of the considered model is written as

$$y_{u,r} = a_u + \beta_r + \tau_{d(u,r)} + \rho_{d(u,r-1)} + \tilde{e}_{u,r} \tag{M0}$$

In this model:

a_u is the (random) effect of unit u;

β_r is the (fixed) effect of period r;

$\tau_{d(u,r)}$ is the (fixed) direct effect of treatment $d(u,r)$;

$\rho_{d(u,r-1)}$ is the (fixed) carryover (left neighbor) effect of the previous treatment $d(u, r-1)$;

$\tilde{e}_{u,r}$ is the (random) error, $1 \leq u \leq n$, $1 \leq r \leq p$.

As can be seen from the indices of $\tau_{d(u,r)}$ and $\rho_{d(u,r-1)}$, these effects are design-dependent. The errors are assumed to be uncorrelated between different units, while errors within a unit u have a covariance $\sigma^2 \Sigma_{du}$. σ^2 is an unknown constant and Σ_{du} is a known, non-singular $p \times p$ matrix. Its off-diagonal entries are 0 or γ, its diagonal elements are equal to 1. A guard plot is not assumed, i.e., $\rho_{d(u,0)} = 0$ as there is no carryover effect influencing the treatment evaluation of the first time period. In contrast to the listed random unit effects, a_u will be treated as fixed in the variance structure of the response since its variance is extremely large.

The covariance matrix Σ_{du}, i.e. the interactions between treatments and subjects, implies the following covariance structure for $y_{u,r}$:

$$Cov(y_{u,r_1}; y_{u,r_2}) = \begin{cases} \sigma^2 & , r_1 = r_2 \\ 0 & , r_1 \neq r_2 \text{ and } d(u,r_1) \neq d(u,r_2) \\ \gamma \sigma^2 & , r_1 \neq r_2 \text{ and } d(u,r_1) = d(u,r_2) \end{cases} \quad (2.1)$$

in which $\gamma \in (0,1)$.

The matrix notation of model (M0) is as follows:

$Y = [y_{1,1}, \ldots, y_{1,p}, y_{2,1}, \ldots, y_{n,p}]^T$ the vector of observations;

$U = I_n \otimes 1_p$ the design matrix of the unit effects;

$P = 1_n \otimes I_p$ the design matrix of the period effects;

$T_d = [T_{d1}^T, \ldots, T_{dn}^T]^T$ the $np \times t$ design matrix of direct treatment effects;

$M_d = [M_{d1}^T, \ldots, M_{dn}^T]^T$ the $np \times t$ design matrix of carryover effects;

$\tilde{E} = [\tilde{e}_{1,1}, \ldots, \tilde{e}_{1,p}, \tilde{e}_{2,1}, \ldots, \tilde{e}_{n,p}]^T$ the vector of errors;

$\Sigma_d = diag(\Sigma_{d1}, \ldots, \Sigma_{dn})$ the covariance matrix of $\tilde{E}$.

The vectors for the unit-, time period-, treatment- and carryover effects are a, β, τ, and ρ, respectively. The vector notation reshapes model (M0) to

$$Y = Ua + P\beta + T_d\tau + M_d\rho + \tilde{E}.$$

In order to apply generalized least squares estimation, it is necessary to transform the response vector Y with a correction matrix V_d. V_d is dependent on the design d. In more detail, let $Cov(\tilde{E})$ be the covariance of the error $\tilde{E}$. Then $Cov(\tilde{E})$ equals $\sigma^2 I_{np} \otimes \Sigma_{du} := \sigma^2 S_d$ and is of dimension $np \times np$. Consider V_d, the $np \times np$ matrix with the properties $V_d V_d = S_d^{-1}$ and $V_d S_d V_d = I_{np}$. Multiplying model (M0) with V_d, the resulting model

$$V_d Y = V_d U a + V_d P \beta + V_d T_d \tau + V_d M_d \rho + E, \tag{M1}$$

has uncorrelated errors because $E = V_d \tilde{E}$ has expectation zero and covariance $\sigma^2 V_d S_d V_d = \sigma^2 I_{np}$.

The information matrix for the least squares estimate of the treatments vector τ occurring in model (M1) is

$$C_d^{(M1)} = T_d^T V_d \omega^\perp (V_d[P, U, M_d]) V_d T_d.$$

The information matrix of a model, which disregards the time periods effect vector β of model (M1), would be

$$C_d = T_d^T V_d \omega^\perp (V_d[U, M_d]) V_d T_d, \tag{2.2}$$

with $C_d^{(M1)} \leq C_d$ in the Loewner-sense. Equality only holds iff $T_d^T V_d \omega^\perp (V_d[U, M_d]) V_d P = 0$. Notice that $T_d 1_t$ is in the column space of U, which implies that C_d has row and column sums zero, see Kunert (1991).

2.2 A Method for Finding Optimal Designs

The main objective of this doctoral thesis is to determine the structure of optimal designs. Therefore, the term of optimality needs to be clarified at first.

2.2.1 The Universal Optimum

In order to judge a design for its optimality, several criteria have been invented. The most common ones are the E-, D- or A- criteria. Extracting the main results from Kiefer (1975), a universally optimal design $d^* \in \Omega_{t,n,p}$ is, among other criteria, optimal under the three listed optimality criteria as well. A design d^* is universally optimal iff its information matrix C_{d^*} is completely symmetric and $\operatorname{tr} C_{d^*}$ is maximal over $\Omega_{t,n,p}$. A matrix M is completely symmetric iff its diagonal elements are identical and its off-diagonal elements coincide as well.

Applying Kiefer (1975) to model (M1), an optimal design d^* is detected through determination of $\max_d [\operatorname{tr} C_d^{(M1)}] = \operatorname{tr} C_{d^*}^{(M1)}$. Because C_d of equation (2.2) is an upper bound for $C_d^{(M1)}$, it is sufficient to determine $\max_d [\operatorname{tr} C_d] = \operatorname{tr} C_{d^*}^{(M1)}$.

2 The Crossover Design and an Instrument for Finding Optimal Designs

2.2.2 Maximizing the Trace of C_d

The intention of finding an optimal design d^* means extensive work if all designs $d \in \Omega_{t,n,p}$ have to be identified. As the formula of C_d depends on the design d, it is difficult to determine $\text{tr}\, C_d$ for an arbitrary design d. Therefore, an upper bound of $\text{tr}\, C_d$ is needed.

Kushner (1997) introduced a method to convert the problem of finding an optimal design into maximizing the minimum of a design-dependent function q_d. Then, this specific minimum of q_d represents an upper bound of $\text{tr}\, C_d$. In order to obtain q_d, the trace of the information matrix C_d needs to be decomposed.

As in Stufken (1996), C_d can be decomposed into

$$C_d = C_{d11} - C_{d12} C_{d22}^- C_{d12}^T,$$

by applying the formula $\omega^\perp([A,B]) = \omega^\perp(A) - \omega^\perp(A)B\left\{B^T\omega^\perp(A)B\right\}^- B^T\omega^\perp(A)$ to C_d, see Math Analogy of Appendix A. The partitions C_{dij}, $1 \leq i,j \leq 2$, are determined by $X_{di}^T \omega^\perp(U) X_{dj}$ with $X_{d1} = T_d$ and $X_{d2} = M_d$. As suggested by Kunert & Martin (2000b), the C_{dij}, $1 \leq i,j \leq 2$, of C_d of equation (2.2) can be transformed into

$$C_{d11} = T_d^T V_d^* T_d,$$
$$C_{d12} = T_d^T V_d^* M_d,$$
$$C_{d22} = M_d^T V_d^* M_d,$$

by defining $V_d^* = diag(V_{d1}^*, \ldots, V_{dn}^*)$ and setting

$$V_{du}^* = V_{du} \omega^\perp (V_{du} 1_p) V_{du} = S_{du}^{-1} - \underbrace{(1_p^T S_{du}^{-1} 1_p)^{-1}}_{\in \mathbb{R}} S_{du}^{-1} 1_p 1_p^T S_{du}^{-1}, \quad (2.3)$$

for all $u = 1, \ldots, n$.

Multiplying the centralizing matrix B_t with the partitions of C_d, yields numerical values $c_{dij} = \text{tr}(B_t C_{dij} B_t)$, for $1 \leq i,j \leq 2$. The matrix $\begin{bmatrix} B_t C_{d11} B_t & B_t C_{d12} B_t \\ B_t C_{d12}^T B_t & B_t C_{d22} B_t \end{bmatrix}$ is nonnegative definite. Using Theorem A.74 of Rao & Toutenburg (1995), the matrix $\begin{bmatrix} c_{d11} & c_{d12} \\ c_{d12} & c_{d22} \end{bmatrix}$ is nonnegative definite as well, which implies that $c_{d11}, c_{d22} \geq 0$ and $c_{d11} c_{d22} - c_{d12}^2 \geq 0$. Thus, we can define a specific design-dependent value

$$q_d^* = c_{d11} \quad \text{, for } c_{d22} = 0 \quad \text{and}$$
$$q_d^* = c_{d11} - c_{d12}^2 / c_{d22} \quad \text{, for } c_{d22} > 0.$$

8

2.2 A Method for Finding Optimal Designs

Proposition 2 of Kunert & Martin (2000a) identifies q_d^* to be an upper bound for $\operatorname{tr} C_d$ that applies for every design $d \in \Omega_{t,n,p}$. Equality holds iff all partitions C_{dij}, $1 \leq i,j \leq 2$, are completely symmetric (and therefore C_d is completely symmetric). Hence, a design is needed, for which this equality is valid for a maximal value of q_d^*. A design f is desirable whose partitioned matrices C_{fij}, $1 \leq i,j \leq 2$, are completely symmetric and $\operatorname{tr} C_f$ is maximal with $C_f = C_f^{(M1)}$. If, additionally, the proportion of units assigned to the treatment sequences yield a maximal q_f^*, then f is optimal for all designs in $\Omega_{t,n,p}$. Unfortunately, there is no guarantee for such a design to exist. But for any design d, q_f^* can still be seen as the best reachable value.

Now, $T_d = [T_{d1}^T, \ldots, T_{du}^T]^T$ and $M_d = [M_{d1}^T, \ldots, M_{du}^T]^T$, i.e., T_{du} and M_{du} are the design matrices for the direct treatment and carryover effects of subject u, $u = 1, \ldots, n$, respectively. Using

$$\begin{aligned} c_{d11}^{(u)} &= \operatorname{tr}\left(B_t(T_{du}^T V_{du}^* T_{du})\right), \\ c_{d12}^{(u)} &= \operatorname{tr}\left(B_t(T_{du}^T V_{du}^* M_{du})\right), \\ c_{d22}^{(u)} &= \operatorname{tr}\left(B_t(M_{du}^T V_{du}^* M_{du})\right), \end{aligned} \qquad (2.4)$$

and equation (2.3), c_{dij} can be written as $\sum_{u=1}^{n} c_{dij}^{(u)}$, for $1 \leq i,j \leq 2$. The $c_{dij}^{(u)}$ are determined by the sequence of the treatments applied to subject u.

Definition 1 (Sequence). A sequence is the prescribed order of treatment samples given to one unit. There is one treatment sample for each time period. The length of a sequence is p.

Definition 2 (Equivalence of Sequences). Two sequences are said to be equivalent iff one is derivable from the other, by relabelling treatments 1 through t appropriately.

Obviously, two units with equivalent sequences have the same values $c_{dij}^{(u)}$, $1 \leq i,j \leq 2$. Hence, for a given t and p, the set of all possible treatment sequences can be grouped according to their equivalence. Define equivalence classes of sequences, $l = 1, \ldots, K$, such that all $c_{dij}^{(u)}$, $1 \leq i,j \leq 2$, are the same for every u receiving a sequence from one class. For example, if $p = 4$, there exist $K = 15$ equivalence classes, given the equivalence of sequences of Definition 2.

Let π_{dl} be the proportion of subjects receiving sequences from the lth-class in a given design $d \in \Omega_{t,n,p}$. Also, define $c_{ij}(l) = c_{dij}^{(u_l)}$, in which u_l is any subject receiving a sequence from the lth class. Then

$$c_{dij} = n \left(\sum_{l=1}^{K} \pi_{dl} c_{ij}(l) \right), \text{ for } 1 \leq i,j \leq 2. \qquad (2.5)$$

2 The Crossover Design and an Instrument for Finding Optimal Designs

Because q_d^* can be calculated from the c_{dij}, the π_{dl} determine q_d^*. But q_d^* is a nonlinear function of the proportions π_{dl}, making maximization of q_d^* through the determination of optimal weights π_{dl} difficult. According to Kunert & Martin (2000a, Proposition 3), this problem can be linearized by using the following proposition.

Proposition 1. For any design $d \in \Omega_{t,n,p}$, define the function $q_d : \mathbb{R} \to \mathbb{R}$ by

$$q_d(x) = c_{d11} + 2xc_{d12} + x^2 c_{d22}. \tag{2.6}$$

Then for all x, we have $q_d(x) \geq q_d^*$, and there is at least one x_d such that $q_d(x_d) = q_d^*$.

Proof. Case 1: $q_d(x) = c_{d11} + 2xc_{d12} + x^2 c_{d22}, \quad c_{d22} > 0.$

By transforming $u = x + c_{d12}/c_{d22}$ or equivalently $x = u - c_{d12}/c_{d22}$, we get

$$q_d(x) = c_{d11} + 2uc_{d12} - 2c_{d12}^2/c_{d22} + (u - c_{d12}/c_{d22})^2 c_{d22}$$
$$= c_{d11} - c_{d12}^2/c_{d22} + u^2 c_{d22} = q_d^* + u^2 c_{d22},$$

which is minimal iff $u^2 c_{d22} = 0$, i.e. $u = 0$. Therefore, $q_d(x) \geq q_d^*$.

Case 2: $\quad q_d(x) = c_{d11} = q_d^*, \quad c_{d12} = 0$ and $c_{d22} = 0.$

Case 3: $\quad q_d(x) = c_{d11} + 2xc_{d12}, \quad c_{d12} \neq 0$ and $c_{d22} = 0.$

Since $c_{d22} = 0$ and $c_{d11}c_{d22} - c_{d12}^2 \geq 0$, we get $c_{d12} = 0$. Thus, we have $q_d(x) = c_{d11}$. It follows: $q_d(x) \geq q_d^*$.

$\square$

Addendum to Proposition 1: The derivative $\frac{\partial q_d(x)}{\partial x}$ in x_d for which $q_d(x_d) = q_d^*$ must be zero, since $q_d(x)$ is a convex function of x.

For each equivalence class $l = 1, \ldots, K$, the h_l function is defined as

$$h_l(x) = c_{11}(l) + 2xc_{12}(l) + x^2 c_{22}(l). \tag{2.7}$$

Apply equations (2.7) and (2.5) to (2.6), and $q_d(x)$ can be written as a linear combination of the h_l functions:

$$q_d(x) = n \sum_{l=1}^{K} \pi_{dl} h_l(x). \tag{2.8}$$

Definition 3 (Approximate Design). Let π_{dl}, $\sum_{l=1}^{k} \pi_{dl} = 1$, be the proportions of subjects which receive sequences from the lth sequence class in a design d. A design $d^* \in \Omega_{t,n,p}$ is said to be an approximate design iff there exists a $n\pi_{dl} \notin \mathbb{N}$, $l = 1, \ldots, k$.

2.2 A Method for Finding Optimal Designs

Taking Proposition 1 of Kunert & Stufken (2002) for the one dimensional q_d function into account, gives

Proposition 2. For a (approximate) design $d^* \in \Omega_{t,n,p}$ consider x_{d^*} for which $q_{d^*}(x_{d^*}) = q_{d^*}^*$. If $nh_l(x_{d^*}) \leq q_{d^*}(x_{d^*}) = q_{d^*}^*$ for all $1 \leq l \leq K$, then $\operatorname{tr} C_d \leq q_{d^*}^*$ for every $d \in \Omega_{t,n,p}$.

Proof. Equivalent to the proof of Proposition 4 of Kunert & Martin (2000a), just replace (x_d, y_d) with (x_{d^*}). □

For the design d^* of Proposition 2, we have $q_{d^*}(x_{d^*})/n = \max_l h_l(x_{d^*})$, which implies

$$q_{d^*}(x_{d^*}) = \min_x q_{d^*}(x) = n \min_x \max_l h_l(x). \tag{2.9}$$

If no design such as d^* of Proposition 2 can be found, the right-hand side of (2.9) is still an upper bound for $\operatorname{tr} C_d$, for any design $d \in \Omega_{t,n,p}$. Figures 2.1 and 2.2 illustrate equations (2.6) and (2.9). The point x_{d^*} of Proposition 2, at which the $\min_x \max_l h_l$ is being realized, lies either at the intersection of two ore more of the h_l, or at the minimum of one h_l function. The function $q_d(x)$ is a linear combination of the h_l forming $\min_x \max_l h_l$.

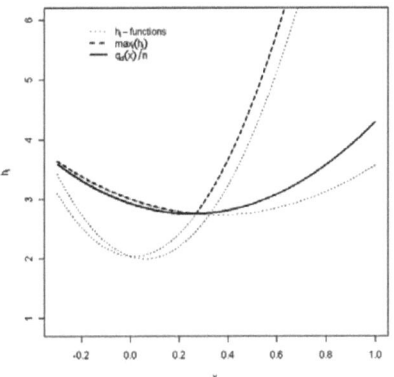

Fig. 2.1: $q_d(x)$ is a linear combination of two h_l functions.

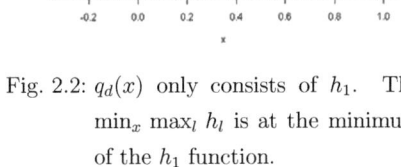

Fig. 2.2: $q_d(x)$ only consists of h_1. The $\min_x \max_l h_l$ is at the minimum of the h_1 function.

Summarizing this chapter, the main intention and, thus, the difficulty consists in the determination of $\min_x \max_l h_l(x)$ for a general number of treatments t and a (general) p. Although some results can be achieved, the generalization of p is beyond the scope of this thesis. Therefore, the focus is put on the usual lengths of time periods $3, 4, 5$ and 6 since they are the most practical sequence lengths used in experiments of, e.g., sensory studies. All parameters are examined with the restriction that $p \leq t$.

3 Deriving General Formulas and Some of Their Properties

Before any extrema of the $h_l(x)$ can be determined, the coefficients $c_{ij}(l)$ for each equivalent class l have to be calculated. In order to simplify equations, some definitions are helpful.

3.1 Definitions of Variables

For any subject u_l receiving a sequence from the lth equivalence class, the following quantities are defined:

t_l the number of (different) treatments in the sequence $(t_l \leq t)$;

$n_j(l)$ the number of treatment j occurring in the sequence, $1 \leq j \leq t_l$;

$\tilde{n}_j(l)$ the number of the carryover effect j occurring in the sequence, i.e., the number of appearances of treatment j followed by any other treatment (including itself);

$\tilde{n}_{ij}(l)$ the number of appearances of treatment j following treatment i in the sequence, whereas $\tilde{n}_{jj}(l)$ is the number of appearances of treatment j following itself (self-carryover effect);

$\tilde{n}_{0j}(l)$ = 1, if treatment j is in the first period; 0 otherwise.

As $c_{ij}(l) = c_{dij}^{(u_l)} \stackrel{(2.4)}{=} \operatorname{tr} B_t C_{dij}^{(u)}$, $1 \leq i,j \leq 2$, the $c_{ij}(l)$ are calculated with the help of the design dependent matrix S_{du}^{-1}, $u = 1, \ldots, n$, which is

$$S_{du}^{-1} = \begin{pmatrix} a_{11} & b_{12} & \cdots & b_{1p} \\ b_{21} & a_{22} & \ddots & \vdots \\ \vdots & \ddots & \ddots & b_{(p-1)p} \\ b_{p1} & \cdots & b_{p(p-1)} & a_{pp} \end{pmatrix}.$$

Matrix entries $a_{rr} \in \{a_{n_j(l)}\}$ and $b_{rr'} \in \{b_{n_j(l)}\}$ for $j = 1, \ldots, t$. Dependent on the treatment appearances $n_j(l)$, the numbers $a_{n_j(l)}$ and $b_{n_j(l)}$ are listed in Table 3.1 for $n_j(l) = 1, \ldots, 6$.

3 Deriving General Formulas and Some of Their Properties

Generally, for $n_j(l) \geq 2$ and $\gamma \in (0,1)$, the formulas of $a_{n_j(l)}$ and $b_{n_j(l)}$ are given as

$$a_{n_j(l)} = \frac{(n_j(l)-2)\gamma + 1}{[(n_j(l)-1)\gamma + 1](1-\gamma)} \quad \text{and}$$

$$b_{n_j(l)} = \frac{-\gamma}{[(n_j(l)-1)\gamma + 1](1-\gamma)}.$$

Note that the parameter γ is the same coefficient as in the covariance structure of $y_{u,r}$, cf. equation (2.1). If treatments in any two or more periods $r, r' = 1, \ldots, p$ coincide, then their coefficients a_{rr} and $a_{r'r'}$ are identical and their coefficients $b_{rr'}$ and $b_{r'r}$ as well. The rth column sum of S_{du}^{-1} equals the rth row sum of S_{du}^{-1} and can be written as

$$crs_{n_{j(r)}(l)} = a_{n_{j(r)}(l)} + (n_{j(r)}(l) - 1)b_{n_{j(r)}(l)},$$

in which $j(r)$ is the treatment j in period r of sequence l. The sum of all matrix entries of S_{du}^{-1} is symbolized by R_u, i.e., $R_u = \sum_{r=1}^{p} crs_{n_{j(r)}(l)}$.

$n_j(l)$	$a_{n_j(l)}$	$b_{n_j(l)}$
1	1	0
2	$1/((1+\gamma)(1-\gamma))$	$-\gamma/((1+\gamma)(1-\gamma))$
3	$(\gamma+1)/((2\gamma+1)(1-\gamma))$	$-\gamma/((2\gamma+1)(1-\gamma))$
4	$(2\gamma+1)/((3\gamma+1)(1-\gamma))$	$-\gamma/((3\gamma+1)(1-\gamma))$
5	$(3\gamma+1)/((4\gamma+1)(1-\gamma))$	$-\gamma/((4\gamma+1)(1-\gamma))$
6	$(4\gamma+1)/((5\gamma+1)(1-\gamma))$	$-\gamma/((5\gamma+1)(1-\gamma))$

Table 3.1: Matrix entries of S_{du}^{-1}, dependent on the number of treatment-j-appearances $n_j(l)$.

Useful term characteristics for discussing properties of $c_{ij}(l)$, $1 \leq i,j \leq 2$, are given by

- $1 = a_1 < a_2 < \cdots < a_p$;

- $b_2 < b_3 < \cdots < b_p < b_1 = 0$;

- $1 = crs_1 > crs_2 > \cdots > crs_p$;

- $1 = crs_1 < 2crs_2 < \cdots < pcrs_p$;

- R_u is increasing as the number of different treatments in sequence u increases, i.e., R_u is maximal if $crs_{n_j(l)}$ is maximal $(= 1)$ $\forall j = 1, \ldots, p$.

The proofs are presented in Appendix C, Lemmas 13 through 16.

3.2 Deriving $c_{11}(l), c_{12}(l)$ and $c_{22}(l)$

Using the definition of $c_{ij}(l) = c_{dij}^{(u_l)}$ and equations (2.4), all three coefficients $c_{ij}(l)$, $1 \leq i,j \leq 2$, can be derived from the design matrices T_d and M_d. Let T_{du}, M_{du} and S_{du}^{-1} be the partial matrices of the design-dependent matrices T_d, M_d and S_d^{-1} for any unit u_l of an equivalence class $l = 1, \ldots, K$. Each row of matrix T_{du} contains exactly one 1 and $(p-1)$ zeros. Thus, the entries of the diagonal matrix $T_{du}^T T_{du}$ equal the number of appearances of treatment j in the sequence of unit u_l. The matrices S_{du}^{-1} and T_{du} are of a structure, in which each element of S_{du}^{-1} that is different from 0 strikes one 1 of T_{du} by multiplying S_{du}^{-1} with T_{du}. The expression $1_t^T \left(T_{du}^T S_{du}^{-1} - \frac{1}{R_u} T_{du}^T S_{du}^{-1} 1_p 1_p^T S_{du}^{-1} \right)$ equals 0, because $1_t^T T_{du}^T = 1_p^T$ and $R_u = 1_p^T S_{du}^{-1} 1_p$ and, therefore, multiplication with B_t could have been omitted in the definition of equations $c_{d11}^{(u)}$ and $c_{d12}^{(u)}$ of (2.4). Analyzing $T_{du}^T S_{du}^{-1} T_{du}$, we see that the entries of the diagonal matrix $T_{du}^T T_{du}$ are additionally multiplied with the corresponding row sum of S_{du}^{-1}. Thus, we obtain

$$\mathrm{tr}\left(T_{du}^T S_{du}^{-1} T_{du}\right) = \sum_{j=1}^{t} n_j(l) crs_{n_j(l)} = R_u, \tag{3.1}$$

the summation of all entries of matrix S_{du}^{-1}, cf. equation (C.1) of the appendix as well. Furthermore, $1_p = T_{du} 1_t$, which forces $T_{du}^T S_{du}^{-1} 1_p$ to be equal to $T_{du}^T S_{du}^{-1} T_{du} 1_t$, the column sums of $T_{du}^T S_{du}^{-1} T_{du}$. The matrix $T_{du}^T S_{du}^{-1} T_{du}$ is diagonal, therefore,

$$\mathrm{tr}\left(T_{du}^T S_{du}^{-1} 1_p R_u^{-1} 1_p^T S_{du}^{-1} T_{du}\right) = \frac{1}{R_u} \sum_{j=1}^{t} \left(n_j(l) crs_{n_j(l)}\right)^2. \tag{3.2}$$

M_{du} is generated by deleting the last row of T_{du} and adding a first row of 0s. Because of this "T_{du}-shifting" structure, the elements of the diagonal matrix $T_{du}^T M_{du}$ equal the number of appearances of the self-carryover effect j $(= \tilde{n}_{jj}(l))$, since M_{du} is the matrix of the carryover effects j in unit u, $j = 1, \ldots, t$. The entries of the diagonal elements of matrix $T_{du}^T S_{du}^{-1} M_{du}$ are, again, multiples of the row sums of S_{du}^{-1}, such that

$$\mathrm{tr}\left(T_{du}^T S_{du}^{-1} M_{du}\right) = \sum_{j=1}^{t} \tilde{n}_{jj}(l) crs_{n_j(l)}. \tag{3.3}$$

The vector $1_p^T S_{du}^{-1} M_{du}$ turns out to be $1_t^T T_{du}^T S_{du}^{-1} M_{du}$, which are the column sums of $T_{du}^T S_{du}^{-1} M_{du}$. The expression $T_{du}^T S_d^{-1} 1_p$ represents the column sums of $T_{du}^T S_{du}^{-1} T_{du}$, and, furthermore,

$$\mathrm{tr}\left(T_{du}^T S_{du}^{-1} 1_p R_u^{-1} 1_p^T S_{du}^{-1} M_{du}\right) = \frac{1}{R_u} \sum_{j=1}^{t} \left(n_j(l) crs_{n_j(l)} \sum_{i=1}^{t} \tilde{n}_{ji}(l) crs_{n_i(l)}\right). \tag{3.4}$$

3 Deriving General Formulas and Some of Their Properties

The summands of the trace of $M_{du}^T M_{du}$ contain all $\tilde{n}_j(l)$, $j = 1, \ldots, t$. The matrix S_{du}^{-1} is not of the "T_{du}-shifted" structure as M_{du}. Thus, the entries of the diagonal matrix $M_{du}^T S_{du}^{-1} M_{du}$ are not exact multiples of the $crs_{n_j(l)}$. They depend on $\tilde{n}_{ij}(l)$, the number of appearances of treatment j following treatment i in sequence l:

$$\text{tr}\left(M_{du}^T S_{du}^{-1} M_{du}\right) = \sum_{i=1}^{t}\sum_{j=1}^{t} \tilde{n}_{ij}(l)\left(a_{n_j(l)} + [\tilde{n}_{ij}(l) - 1]b_{n_j(l)}\right). \tag{3.5}$$

Similar to T_{du}, the matrix M_{du} contains exactly one 1 in every but the first row. Hence, $1_t^T M_{du}^T = (0, 1, \ldots, 1)$ is a row vector with its first entry equal to 0 and the following $(p-1)$ entries equal to 1. Multiplication of the matrix $1_t 1_t^T$ with $M_{du}^T S_{du}^{-1} M_{du}$ leads to a trace, in which all entries of S_{du}^{-1} are added except the first row and column, i.e.,

$$\text{tr}\left(1_t 1_t^T M_{du}^T S_{du}^{-1} M_{du}\right) = \sum_{j=1}^{t} (n_j(l) - \tilde{n}_{0j}(l))\left(crs_{n_j(l)} - \tilde{n}_{0j}(l)b_{n_j(l)}\right). \tag{3.6}$$

Similar to equation (3.2), and using the annotations for deriving equation (3.4), we get

$$\text{tr}\left(M_{du}^T S_{du}^{-1} 1_p R_u^{-1} 1_p^T S_{du}^{-1} M_{du}\right) = \frac{1}{R_u} \sum_{i=1}^{t}\left(\sum_{j=1}^{t} \tilde{n}_{ij}(l) crs_{n_j(l)}\right)^2. \tag{3.7}$$

The vectors 1_p and $M_{du} 1_t$ differ only in its first element. Thus, $1_t^T M_{du}^T S_{du}^{-1} 1_p$ is the sum of all elements of the matrix S_{du}^{-1}, except the first column. However, the first row elements of $1_t^T M_{du}^T S_{du}^{-1} M_{du} 1_t$ have been left out as well, cf. the derivation of equation (3.6). Therefore,

$$\text{tr}\left(1_t 1_t^T M_{du}^T S_{du}^{-1} 1_p R_u^{-1} 1_p^T S_{du}^{-1} M_{du}\right) = \frac{1}{R_u}\left(\sum_{j=1}^{t}(n_j(l) - \tilde{n}_{0j}(l))crs_{n_j(l)}\right)^2. \tag{3.8}$$

Inserting equations (3.1) through (3.8) into equations (2.4), it is easy to derive that

$$c_{11}(l) = R_u - \frac{1}{R_u} \sum_{j=1}^{t} n_j^2(l) crs_{n_j(l)}^2, \qquad (3.9)$$

$$c_{12}(l) = \sum_{j=1}^{t} \tilde{n}_{jj}(l) crs_{n_j(l)} - \frac{1}{R_u} \sum_{j=1}^{t} \left(n_j(l) crs_{n_j(l)} \sum_{i=1}^{t} \tilde{n}_{ji}(l) crs_{n_i(l)} \right) \qquad (3.10)$$

and

$$\begin{aligned} c_{22}(l) = &\sum_{i=1}^{t} \sum_{j=1}^{t} \tilde{n}_{ij}(l) \left(a_{n_j(l)} + [\tilde{n}_{ij}(l) - 1] b_{n_j(l)} \right) \\ &- \frac{1}{t} \sum_{j=1}^{t} (n_j(l) - \tilde{n}_{0j}(l)) \left(crs_{n_j(l)} - \tilde{n}_{0j}(l) b_{n_j(l)} \right) \\ &- \frac{1}{R_u} \sum_{i=1}^{t} \left(\sum_{j=1}^{t} \tilde{n}_{ij}(l) crs_{n_j(l)} \right)^2 \\ &+ \frac{1}{t R_u} \left(\sum_{j=1}^{t} (n_j(l) - \tilde{n}_{0j}(l)) crs_{n_j(l)} \right)^2 . \end{aligned} \qquad (3.11)$$

Using some transformations, the second and fourth term of $c_{22}(l)$ above can be simplified to $\frac{1}{tR_u} \left(\sum_j \tilde{n}_{0j(l)} crs_{n_j(l)} \right)^2 - \frac{1}{t} \sum_j \tilde{n}_{0j}(l) a_{n_j(l)}$, which provides an equivalent expression:

$$\begin{aligned} c_{22}(l) = &\sum_{i=1}^{t} \sum_{j=1}^{t} \tilde{n}_{ij}(l) \left(a_{n_j(l)} + [\tilde{n}_{ij}(l) - 1] b_{n_j(l)} \right) - \frac{1}{t} \sum_{j=1}^{t} \tilde{n}_{0j}(l) a_{n_j(l)} \\ &- \frac{1}{R_u} \sum_{i=1}^{t} \left(\sum_{j=1}^{t} \tilde{n}_{ij}(l) crs_{n_j(l)} \right)^2 + \frac{1}{t R_u} \left(\sum_{j=1}^{t} \tilde{n}_{0j}(l) crs_{n_j(l)} \right)^2 . \end{aligned} \qquad (3.12)$$

An illustration of the mentioned transformation is presented in section C.2 of Appendix C.

3.3 Some Properties of $c_{11}(l)$ and $c_{12}(l)$

Some simplifications of formulas and properties, concerning $n_j(l)$-dependent behavior of the $c_{ij}(l)$ functions, are gathered in this section.

Definition 4 (In-/ and Decreasing $\mathbf{n_j(l)}$ of Sequence $\mathbf{u_l}$). Let u_l and $u_{l'}$ be two sequences, in which all treatments except for k and j appear equally often. Suppose, $n_k(l) < n_k(l')$ and $n_j(l) > n_j(l')$. Further, assume that there is an $x \geq 1$, such that $n_k(l') = n_k(l) + x$ and $n_j(l') = n_j(l) - x$.

3 Deriving General Formulas and Some of Their Properties

The number of appearances of treatment k is said to be increasing in $u_{l'}$ iff $n_k(l) \geq n_j(l)$, $n_k(l') \leq p$ and $n_j(l') \geq 0$. Alternatively, the number of appearances of treatment j is said to be decreasing in $u_{l'}$ iff $n_j(l) \geq n_k(l)$, $n_k(l') \leq p$ and $n_j(l') > n_k(l)$.

Illustration of Definition 4:

$$\text{sequence } u_l: \qquad [j\,j\,2\,3\,4\,k\,j]$$

$$n_j(l) \text{ is increasing} \uparrow \qquad \downarrow n_j(l) \text{ is decreasing}$$

$$\text{sequence } u_{l'}: \qquad [j\,j\,3\,2\,k\,k\,4]$$

Notice, the number of different treatments $t_{l'}$ in sequence $u_{l'}$ decreases iff $n_j(l') \searrow 0$ for any $j = 1, \ldots, t_{l'}$ and it increases iff a new treatment j is added to sequence $u_{l'}$ such that $n_j(l') > 0$ for any $j = \{1, \ldots, t\} \setminus \{1, \ldots, t_l\}$.

Lemma 1. In a sequence u_l of class l, $c_{11}(l)$ decreases if $n_j(l)$ increases for any treatment $j = 1, \ldots, t_l$, in the sense of Definition 4.

Proof.

$$c_{11}(l) = R_u \left(1 - \frac{\sum_{j=1}^{t} n_j^2(l) crs_{n_j(l)}^2}{\left(\sum_{j=1}^{t} n_j(l) crs_{n_j(l)}\right)^2}\right) := R_u \left(1 - \frac{\sum_{j=1}^{t} \eta_j^2}{\left(\sum_{j=1}^{t} \eta_j\right)^2}\right).$$

(1) $\eta_j \geq 1$ for all j and

$$\sum_j \eta_j^2 \leq \left(\sum_j \eta_j\right)^2 \quad \forall \eta_j. \tag{a}$$

Set $t = t_l$ since $n_j(l) = 0$ for all treatments j being not in sequence u_l, such that $\sum_j^{t} n_j(l) crs_{n_j(l)} = \sum_j^{t_l} n_j(l) crs_{n_j(l)}$. Further, write

$$\left(\sum_j^{t} \eta_j\right)^2 = \sum_{j=1}^{t} \eta_j^2 + 2\sum_{i=1}^{t-1} \sum_{j>i}^{t} \eta_i \eta_j. \tag{b}$$

Using equation (b), we see that $x_{ratio} := \dfrac{\sum_{j=1}^{t} \eta_j^2}{\left(\sum_{j=1}^{t} \eta_j\right)^2}$ is equal to $\dfrac{\sum_{j=1}^{t} \eta_j^2}{\sum_{j=1}^{t} \eta_j^2 + 2\sum_{i=1}^{t-1} \sum_{j>i}^{t} \eta_i \eta_j}$ and is in the interval $(0,1)$, cf. equation (a). The magnitude of x_{ratio} is determined by expression $2\sum_{i=1}^{t-1} \sum_{j>i}^{t} \eta_i \eta_j$, i.e., iff $2\sum_{i=1}^{t-1} \sum_{j>i}^{t} \eta_i \eta_j$ increases, x_{ratio} decreases and vice versa.

Thus, $\left(\sum_{j}^{t} \eta_j\right)^2$ consists of t summands η_j^2 and $t(t-1)$ summands $\eta_i \eta_j$, $i \neq j$, $i,j = 1,\ldots,t$.

Let further be $n_1(l) \leq n_2(l) \leq \cdots \leq n_t(l)$, then $\eta_1(l) \leq \eta_2(l) \leq \cdots \leq \eta_t(l)$, cf. the listed characteristics of section 3.1. In general, $2\sum_i \sum_{j>i} \eta_i \eta_j$ has the following structure

$$2\sum_i \sum_{j>i} \eta_i \eta_j = 2\eta_1 \eta_2 + 2\eta_1 \eta_3 + \ldots + 2\eta_1 \eta_t + \ldots \atop + 2\eta_k \eta_{k+1} + \ldots + 2\eta_k \eta_t + \ldots + 2\eta_{t-1}\eta_t. \tag{c0}$$

Without loss of generality, an arbitrary n_k, $1 < k \leq t$, decreases to $n'_k = n_k - x$, $x \in \mathbb{N}$, implying that n_1 increases to $n'_1 = n_1 + x$. Thus, $2\sum_i \sum_{j>i} \eta_i \eta_j$ becomes

$$2\sum_i \sum_{j>i} \eta'_i \eta'_j = 2\eta'_1 \eta_2 + 2\eta'_1 \eta_3 + \ldots + 2\eta'_1 \eta'_k + 2\eta'_1 \eta_{k+1} + \ldots + 2\eta'_1 \eta_t + \atop + \ldots + 2\eta'_k \eta_{k+1} + \ldots + 2\eta'_k \eta_t + \ldots + 2\eta_{t-1}\eta_t, \tag{c1}$$

with the assumption that $n'_k = n_k - x \geq n_1$. This assumption is equivalent to $n'_1 = n_1 + x \leq n_k$, see Definition 4. The difference of (c0) and (c1) has to be positive, i.e.,

$$2\sum_i \sum_{j>i} \eta'_i \eta'_j - 2\sum_i \sum_{j>i} \eta_i \eta_j \overset{!}{>} 0$$
$$\Leftrightarrow \eta_2(\eta'_1 - \eta_1) + \eta_3(\eta'_1 - \eta_1) + \ldots + \eta'_k \eta'_1 - \eta_k \eta_1 + \eta_{k+1}(\eta'_1 - \eta_1) + \ldots$$
$$+ \eta_t(\eta'_1 - \eta_1) + (\eta'_k - \eta_k)\eta_{k+1} + \ldots + (\eta'_k - \eta_k)\eta_t \overset{!}{>} 0$$
$$\Leftrightarrow \eta_2 x(1-\gamma) crs_{n_1} crs_{n'_1} + \eta_3 x(1-\gamma) crs_{n_1} crs_{n'_1} + \ldots + (\eta'_k \eta'_1 - \eta_k \eta_1) + \tag{d}$$
$$+ x(1-\gamma)(\eta_{k+1} + \ldots + \eta_t) \underbrace{(crs_{n_1} crs_{n'_1} - crs_{n_k} crs_{n'_k})}_{:=(*)>0} \overset{!}{>} 0$$
$$\Leftrightarrow \eta_2 x(1-\gamma) crs_{n_1} crs_{n'_1} + \eta_3 x(1-\gamma) crs_{n_1} crs_{n'_1} + \ldots + (\eta'_k \eta'_1 - \eta_k \eta_1) \overset{!}{>} 0.$$

Inequality (d) holds because

$$\eta'_k \eta'_1 - \eta_k \eta_1 = x(1-\gamma) \underbrace{(n_k - n_1 - x)}_{=n_k - n'_1 \geq 0}(n_1 \gamma + n_k \gamma + 1 - \gamma) crs_{n_k} crs_{n_1} crs_{n'_k} crs_{n'_1} > 0,$$

for all $x \in \mathbb{N}$ and $\gamma \in (0,1)$; and, as indicated above, $(*) > 0$ because $n_1 \leq n_k$, $crs_{n_1} > crs_{n'_k}$ and $crs_{n'_1} \geq crs_{n_k}$. The reverse conclusion can be drawn for $-x \in \mathbb{N}$, i.e., some n_k increases. Thus, equation (d) proves that, $2\sum_i \sum_{j>i} \eta_i \eta_j$ increases iff some

3 Deriving General Formulas and Some of Their Properties

$n_k = n_k(l)$ increases.

Therefore, x_{ratio} decreases iff $n_k(l)$ decreases for any treatment $k = 1, \ldots, t_l$. The ratio x_{ratio} approaches its minimum $\frac{1}{\sum_j n_j} \ll 1$, iff $n_j(l) = 1$ for all $j = 1, \ldots, t_l$, such that $t_l = p$.

(2) R_u decreases as $n_j(l)$ increases, cf. Lemma 16, Appendix C.

Lemma 1 follows with properties (1) and (2). □

Lemma 2. Assuming $n_j(l)$ is fixed $\forall j = 1, \ldots, t$, $c_{12}(l)$ is maximized by ordering treatments according to ascending $n_j(l)$ and assigning identical treatments consecutively.

Proof. Let

$$A = \sum_{j=1}^{t} \tilde{n}_{jj}(l) crs_{n_j(l)} \text{ and}$$

$$B = \frac{1}{R_u} \sum_{j=1}^{t} \left(n_j(l) crs_{n_j(l)} \sum_{i=1}^{t} \tilde{n}_{ji}(l) crs_{n_i(l)} \right),$$

then $c_{12}(l) = A - B$. Without loss of generality, assume that $n_1 \leq n_2 \leq \cdots \leq n_t$, in which $t := t_l$ is the number of different treatments in sequence u_l. The purpose is to maximize term A and minimize term B. The assumption is that the $n_j(l)$, or $crs_{n_j(l)}$ respectively, are fixed for all $j = 1, \ldots, t$. Thus, we need to find a special order of treatments that maximizes $c_{12}(l)$.

(1) A is maximal $\Leftrightarrow \tilde{n}_{jj}(l) = n_j(l) - 1$ is maximal $\Leftrightarrow$ identical treatments are ordered consecutively.

(2) The intention is to minimize B, which is equivalent to minimize $R_u B$ as R_u is fixed iff all $n_j(l)$ are fixed. The measure to evaluate if $R_u B$ decreases is the performance of $R_u B$ after certain shifting of treatments.

Therefore, first, consider an arbitrary sequence

$$S_{01} = [\cdots \quad a \quad k \quad b \quad \cdots \quad f \quad m \quad \cdots \ ]$$

with treatments $a, b, f, k, m \in \{1, \ldots, t\}$ and their number of appearances are such that

3.3 Some Properties of $c_{11}(l)$ and $c_{12}(l)$

$n_a, n_b, n_f \leq n_k \leq n_m$. The size of term $R_u B$ of sequence S_{01} is described by

$$R_u B_{01} = \sum_{j}^{\{1...t\}\setminus\{a,f,k\}} \left(n_j crs_{n_j} \sum_{i=1}^{t} \tilde{n}_{ji} crs_{n_i} \right)$$

$$+ n_a crs_{n_a} \left(\sum_{i}^{\{1...t\}\setminus\{k,b\}} \tilde{n}_{ai} crs_{n_i} + \tilde{n}_{ak} crs_{n_k} + \tilde{n}_{ab} crs_{n_b} \right)$$

$$+ n_f crs_{n_f} \left(\sum_{i}^{\{1...t\}\setminus\{k,m\}} \tilde{n}_{fi} crs_{n_i} + \tilde{n}_{fk} crs_{n_k} + \tilde{n}_{fm} crs_{n_m} \right)$$

$$+ n_k crs_{n_k} \left(\sum_{i}^{\{1...t\}\setminus\{b,m\}} \tilde{n}_{ki} crs_{n_i} + \tilde{n}_{km} crs_{n_m} + \tilde{n}_{kb} crs_{n_b} \right).$$

Based on the structure and assumptions of S_{01}, the first shifting structure of treatments is implemented as follows:

Shifting 1: $[\ \cdots\ a\ \widehat{k\ b}\ \cdots\ \widehat{f\ m}\ \cdots\ ] \cong [\ \cdots\ a\ b\ \cdots\ f\ k\ m\ \cdots\ ].$

Shifting 1 modifies the magnitude of term $R_u B_{01}$ to

$$R_u B_1 = \sum_{j}^{\{1...t\}\setminus\{a,f,k\}} \left(n_j crs_{n_j} \sum_{i=1}^{t} \tilde{n}_{ji} crs_{n_i} \right)$$

$$+ n_a crs_{n_a} \left(\sum_{i}^{\{1...t\}\setminus\{k,b\}} \tilde{n}_{ai} crs_{n_i} + (\tilde{n}_{ak} - 1) crs_{n_k} + (\tilde{n}_{ab} + 1) crs_{n_b} \right)$$

$$+ n_f crs_{n_f} \left(\sum_{i}^{\{1...t\}\setminus\{k,m\}} \tilde{n}_{fi} crs_{n_i} + (\tilde{n}_{fk} + 1) crs_{n_k} + (\tilde{n}_{fm} - 1) crs_{n_m} \right)$$

$$+ n_k crs_{n_k} \left(\sum_{i}^{\{1...t\}\setminus\{b,m\}} \tilde{n}_{ki} crs_{n_i} + (\tilde{n}_{km} + 1) crs_{n_m} + (\tilde{n}_{kb} - 1) crs_{n_b} \right).$$

Analyze now the difference $R_u(B_{01} - B_1)$ to evaluate the performance of $R_u B$ when Shifting 1 is applied to an arbitrary sequence S_{01}. Three cases are required:

Case 1: Treatments $a \neq k$ and k is not originally placed in period $p = 1$.

$$R_u(B_{01} - B_1) = n_a crs_{n_a} \underbrace{(crs_{n_k} - crs_{n_b})}_{\leq 0 \text{ as } n_b \leq n_k} + n_f crs_{n_f} \underbrace{(crs_{n_m} - crs_{n_k})}_{\leq 0 \text{ as } n_b \leq n_k}$$

$$+ n_k crs_{n_k} \underbrace{(crs_{n_b} - crs_{n_m})}_{\geq 0 \text{ as } n_b \leq n_m}.$$

3 Deriving General Formulas and Some of Their Properties

Define $n_*crs_{n_*} = \max\{n_a crs_{n_a}, n_f crs_{n_f}\}$ to obtain

$$R_u(B_{01} - B_1) \geq n_* crs_{n_*}(crs_{n_m} - crs_{n_b}) + n_k crs_{n_k}(crs_{n_b} - crs_{n_m})$$
$$= \underbrace{(n_k crs_{n_k} - n_* crs_{n_*})}_{\geq 0 \text{ as } n_k \geq n_* \in \{n_a, n_f\}}(crs_{n_b} - crs_{n_m})$$
$$\geq 0.$$

Case 2: Treatments $a = k$ and k is not originally placed in period $p = 1$.

The partitioning of the term $R_u B_{01}$ in $n_a crs_{n_a} \sum_{i}^{t} \tilde{n}_{ai} crs_{n_i}$ can be omitted. The same applies for summand $n_k crs_{n_k} \tilde{n}_{kb} crs_{n_b}$, which is replaced by $n_k crs_{n_k} \tilde{n}_{kk} crs_{n_k}$. Similar to $R_u B_{01}$, the same partitioned structure applies to $R_u B_1$. It follows that

$$R_u(B_{01} - B_1) = n_f crs_{n_f} \underbrace{(crs_{n_m} - crs_{n_k})}_{\leq 0 \text{ as } n_b \leq n_k} + n_k crs_{n_k} \underbrace{(crs_{n_b} - crs_{n_m})}_{\geq 0 \text{ as } n_b \leq n_m}$$
$$\geq \underbrace{(n_k crs_{n_k} - n_f crs_{n_f})}_{\geq 0 \text{ as } n_f \leq n_k} \underbrace{(crs_{n_k} - crs_{n_m})}_{\geq 0 \text{ as } n_k \leq n_m}$$
$$\geq 0.$$

Case 3: Treatment k is originally placed in period $p = 1$ and, thus, k is no successor of any treatment in the sequence.

As in Case 2, the partitioning of terms $R_u B_{01}$ and $R_u B_1$ in summand $n_a crs_{n_a} \sum_{i}^{t} \tilde{n}_{ai} crs_{n_i}$ can be omitted, which causes $R_u(B_{01} - B_1)$ to become

$$R_u(B_{01} - B_1) = n_f crs_{n_f} \underbrace{(crs_{n_m} - crs_{n_k})}_{\leq 0 \text{ as } n_b \leq n_k} + n_k crs_{n_k} \underbrace{(crs_{n_b} - crs_{n_m})}_{\geq 0 \text{ as } n_b \leq n_m}$$
$$\geq n_k crs_{n_k} \underbrace{(crs_{n_b} - crs_{n_k})}_{\geq 0 \text{ as } n_b \leq n_k}$$
$$\geq 0.$$

Hence, $R_u B_1 \leq R_u B_{01}$. The conclusion is that term B is getting smaller by shifting an arbitrary treatment k to a position such that k is followed by a treatment that appears at least k times in the sequence. However, the assumption must be that treatment k is originally placed between two treatments with number of appearances less than n_k. Shifting 1 leaves B unaltered iff $n_b = n_k = n_m$.

The second shifting structure to be implemented is based on an arbitrary sequence

$$S_{02} = [\ \cdots\ a\ t\ b\ \cdots\ f\ ],$$

3.3 Some Properties of $c_{11}(l)$ and $c_{12}(l)$

in which t is the treatment with the largest number of appearances in the sequence, formally, $n_j \leq n_t$ for all $j = 1, \ldots, t-1$. Treatments a, b, f of sequence S_{02} are elements of the set of treatments $\{1, \ldots, t-1\}$. Shifting 2 defines the movement of t to the last period, i.e.,

Shifting 2: $\quad [\; \cdots \; a \; \overbrace{t \; b \; \cdots \; f} \;] \cong [\; \cdots \; a \; b \; \cdots \; f \; t \;]$.

The magnitude of the term $R_u B$ of sequence S_{02} is given by

$$R_u B_{02} = \sum_{j}^{\{1\ldots t\}\setminus\{a,f,t\}} \left(n_j crs_{n_j} \sum_{i=1}^{t} \tilde{n}_{ji} crs_{n_i} \right)$$
$$+ n_a crs_{n_a} \left(\sum_{i}^{\{1\ldots t\}\setminus\{b,t\}} \tilde{n}_{ai} crs_{n_i} + \tilde{n}_{at} crs_{n_t} + \tilde{n}_{ab} crs_{n_b} \right)$$
$$+ n_f crs_{n_f} \left(\sum_{i}^{\{1\ldots t-1\}} \tilde{n}_{fi} crs_{n_i} + \tilde{n}_{ft} crs_{n_t} \right)$$
$$+ n_t crs_{n_t} \left(\sum_{i}^{\{1\ldots t\}\setminus\{b\}} \tilde{n}_{ti} crs_{n_i} + \tilde{n}_{tb} crs_{n_b} \right).$$

After applying Shifting 2 to sequence S_{02}, the expression of $R_u B_{02}$ becomes

$$R_u B_2 = \sum_{j}^{\{1\ldots t\}\setminus\{a,f,t\}} \left(n_j crs_{n_j} \sum_{i=1}^{t} \tilde{n}_{ji} crs_{n_i} \right)$$
$$+ n_a crs_{n_a} \left(\sum_{i}^{\{1\ldots t\}\setminus\{b,t\}} \tilde{n}_{ai} crs_{n_i} + (\tilde{n}_{at} - 1) crs_{n_t} + (\tilde{n}_{ab} + 1) crs_{n_b} \right)$$
$$+ n_f crs_{n_f} \left(\sum_{i}^{\{1\ldots t-1\}} \tilde{n}_{fi} crs_{n_i} + (\tilde{n}_{ft} + 1) crs_{n_t} \right)$$
$$+ n_t crs_{n_t} \left(\sum_{i}^{\{1\ldots t\}\setminus\{b\}} \tilde{n}_{ti} crs_{n_i} + (\tilde{n}_{tb} - 1) crs_{n_b} \right).$$

Let us now analyze the difference of term $R_u B$ before and after Shifting 2, i.e., consider $R_u(B_{02} - B_2)$ for the required cases:

Case 1: treatment $a \neq t$ and t is not originally placed in period $p = 1$.

3 Deriving General Formulas and Some of Their Properties

$$R_u(B_{02} - B_2) = n_a crs_{n_a} \underbrace{(crs_{n_t} - crs_{n_b})}_{\leq 0 \text{ as } n_b \leq n_t} \underbrace{-n_f crs_{n_f} crs_{n_t}}_{\geq n_t crs_{n_t}} + n_t crs_{n_t} crs_{n_b}$$

$$\geq \underbrace{(n_t crs_{n_t} - n_a crs_{n_a})}_{\geq 0 \text{ as } n_t \geq n_a} \underbrace{(crs_{n_b} - crs_{n_t})}_{\geq 0 \text{ as } n_b \leq n_t}$$

$$\geq 0.$$

Case 2: treatment $a = t$ and t is not originally placed in period $p = 1$.
The partitioning of term $R_u B_{02}$ in summand $n_a crs_{n_a} \sum_i^t \tilde{n}_{ai} crs_{n_i}$ can be omitted. The same applies for summand $n_t crs_{n_t} \tilde{n}_{tb} crs_{n_b}$, which is replaced by $n_t crs_{n_t} \tilde{n}_{tt} crs_{n_t}$. Similar to $R_u B_{02}$, term $R_u B_2$ is readjusted, respectively. As a result, we obtain

$$R_u(B_{02} - B_2) = n_t crs_{n_t} crs_{n_t} - n_f crs_{n_f} crs_{n_t}$$

$$= crs_{n_t} \underbrace{(n_t crs_{n_t} - n_f crs_{n_f})}_{\geq 0 \text{ as } n_t \geq n_f}$$

$$\geq 0.$$

Case 3: Treatment t is originally placed in period $p = 1$ and, thus, t is no successor of any treatment in the sequence.
As in Case 2, the partitioning of terms $R_u B_{02}$ and $R_u B_2$ in summand $n_a crs_{n_a} \sum_i^t \tilde{n}_{ai} crs_{n_i}$ can be omitted, which causes $R_u(B_{02} - B_2)$ to take the value

$$R_u(B_{02} - B_2) = n_t crs_{n_t} crs_{n_b} \underbrace{-n_f crs_{n_f}}_{\geq n_t crs_{n_t}} crs_{n_t}$$

$$\geq n_t crs_{n_t} \underbrace{(crs_{n_b} - crs_{n_t})}_{\geq 0 \text{ as } n_b \leq n_t}$$

$$\geq 0.$$

Hence, $R_u B_2 \leq R_u B_{02}$. The conclusion is that term B can be further reduced by shifting the treatment t with the largest number of appearances to the last position in the sequence. Shifting 2 generates a constant value of B iff $n_b = n_f = n_t$.

Taking into account the definition of shifting structures 1 and 2, it is possible to prove that sequences of the form

$$[\underbrace{1 \cdots 1}_{\sharp = n_1}, \underbrace{2 \cdots 2}_{\sharp = n_2}, \ldots, \underbrace{t-1 \cdots t-1}_{\sharp = n_{t-1}}, \underbrace{t \cdots t}_{\sharp = n_t}]$$

generate the maximum possible value of $c_{12}(l)$, assuming that all $n_j(l)$, $j = 1, \ldots, t$, are fixed and $n_1 \leq n_2 \leq \cdots \leq n_{t-1} \leq n_t$.

3.3 Some Properties of $c_{11}(l)$ and $c_{12}(l)$

The initial situation consists of an arbitrary sequence S_{ini}, e.g. symbolized as $S_{ini} = S_{02}$. The first improvement in minimizing B is done by picking any treatment t and moving it to period p, i.e., by application of Shifting 2. The "worst" case would be if the B-values before and after the shifting are identical. Thus, Shifting 2 is applicable in minimizing B, as it does not change the value of B for the worse, see the definition of Shifting 2 above. The present situation is symbolized by the sequence

$$S_1 = [\quad \cdots \quad a \quad b \quad \cdots \quad f \quad t \quad].$$

As t is the treatment with $n_j(l) \leq t$, $\forall j = 1, \ldots, t-1$, Shifting 1 guarantees further minimization of B by applying Shifting 1 $(n_t - 1)$ times to all other treatments t of sequence S_1, which are not placed in period p. This operation converts S_1 to the sequence

$$S_2 = [\quad \cdots \quad , \quad \underbrace{t \cdots t}_{\sharp = n_t} \quad].$$

The number of appearances of the remaining treatments in periods 1 through $p - n_t$ of S_2 are all less or equal to n_{t-1}. However, this implies that the application of Shifting 1 is feasible in minimizing B further. Shifting 1 means to move any treatment $t-1$ to period $p - n_t$. The repetition of Shifting 1, and, thus, improvement of B, to all other $n_{t-1} - 1$ treatments $t-1$ gives sequence

$$S_3 = [\quad \cdots \quad , \quad \underbrace{t-1 \cdots t-1}_{\sharp = n_{t-1}} \quad , \quad \underbrace{t \cdots t}_{\sharp = n_t} \quad],$$

in which treatments $t-1$ are placed into periods $p - n_t$ through $p - n_t - n_{t-1} + 1$. The number of appearances of all remaining treatments in periods 1 through $p - n_t - n_{t-1}$ of sequence S_3 do not exceed n_{t-2}. Hence, in the sense of Shifting 1, the ordering of all consecutively appointed treatments $t-2$ as predecessor of treatment $t-1$ guarantees further minimization of B, since it is equivalent to ordering all treatments $t-1$ consecutively and in front of treatment t. Furthermore, the sequencing application of Shifting 1 to all treatments $t-2$, $t-3$, etc., results in a stepwise minimization of B until, finally, Shifting 1 allocates treatment 1 to be placed to periods $n_1 = p - n_t - n_{t-1} - \cdots - n_2$ through $1 = p - n_t - n_{t-1} - \cdots - n_2 - n_1 + 1$. The final sequence yielding the minimal value of B is

$$[\underbrace{1 \cdots 1}_{\sharp = n_1}, \underbrace{2 \cdots 2}_{\sharp = n_2}, \quad \cdots \quad , \underbrace{t-1 \cdots t-1}_{\sharp = n_{t-1}}, \underbrace{t \cdots t}_{\sharp = n_t}].$$

Properties (1) and (2) are consistent, and Lemma 2 follows. □

3 Deriving General Formulas and Some of Their Properties

3.4 Special Sequence Class Functions h_l

There are two equivalence class functions of sequences, which prove to be crucial for this thesis: The first class, $l = 1$, to be introduced, contains all sequences equivalent to $[1, 2, \ldots, p-1, p]$, i.e., there is a different treatment given to the unit in every single period. The second class, $l = 2$, represents $[1, 2, \ldots, p-1, p-1]$, i.e., there is a different treatment given to the unit in periods 1 through $p - 1$ and the treatment of period $p - 1$ is to be evaluated in period p a second time. Applying equations (3.9), (3.10), (3.12) to the sequence structures given above, and inserting the results into h_l of equation (2.7), the equivalence class functions h_1 and h_2 are given as

$$h_1(x) = p - 1 - \frac{2(p-1)}{p} \cdot x + \frac{(p-1)(tp-t-1)}{tp} \cdot x^2,$$

with $R_1 = p$ and

$$h_2(x) = \frac{(p-2)(p-3)\gamma + p^2 - p - 2}{(p-2)\gamma + p} - \frac{2(p-3)\gamma}{(p-2)\gamma + p} \cdot x$$
$$+ \frac{((p-3)^2 t - (p-3))\gamma^2 + 2((p-3)t - 1)\gamma - (p-1)^2 t + p - 1}{((p-2)\gamma^2 + 2\gamma - p)t} \cdot x^2,$$

with $R_2 = [p + (p-2)\gamma]/(1+\gamma)$.

The sequence class functions h_1 and h_2 intersect once at $x = 1/(p-1)$ if $\gamma = 0$. However, if $\gamma \in (0, 1)$, there are two intersections located at

$$x_{1/2} = \frac{\mp\sqrt{W} - V}{Z}, \tag{3.13}$$

in which

$$W = [(-4p^4 + 16p^3 - 12p^2 - 8p + 4)t^2 + (4p^2 - 8p)t]\gamma^4 + [(4p^4 - 24p^3 + 28p^2 + 16p - 8)t^2 + (20p - 8p^2)t]\gamma^3 + [(p^4 + 6p^3 - 23p^2 - 8p + 4)t^2 + (4p^2 - 16p)t]\gamma^2 + [(-2p^4 + 4p^3 + 6p^2)t^2 + 4pt]\gamma + (p^4 - 2p^3 + p^2)t^2,$$
$$V = tp(p-1)(1-\gamma) + 2t\gamma(1-\gamma)$$

and

$$Z = [(2p^2 - 4p - 2)t - 2]\gamma^2 + [(2p+2)t + 2]\gamma.$$

The value x_1 is smaller than x_2 because the numerator of x_1 contains a negative root, whereas the root of x_2 is positive. The denominator is positive for all $p \geq 3$. Some further equivalent transformations lead to the following proposition.

Proposition 3. Let $\gamma \in (0, 1)$ and $t \geq p \geq 3$. Then, the intersection point, in which $h_1(x) = h_2(x)$ is smallest, is located at $x_2 \in (0, 1)$ of equation (3.13).

3.4 Special Sequence Class Functions h_l

Proof. (1)

$$x_2 > 0 \Leftrightarrow \left(\sqrt{W}\right)^2 \geq (tp(p-1)(1-\gamma) + 2t\gamma(1-\gamma))^2$$
$$\overset{:4pt\gamma}{\Leftrightarrow} (p-2 + (-p^3 + 4p^2 - 3p - 2)t)\gamma^3 + (5 - 2p + (p^3 - 6p^2 + 6p + 5)t)\gamma^2$$
$$+ (p - 4 + 2(p^2 - 2p - 2)t)\gamma + 1 + (p+1)t > 0$$
$$:= g(\gamma) > 0$$

Rewrite $g(\gamma)$ as $g(\gamma) = g_0 + g_1\gamma + g_2\gamma^2 + g_3\gamma^3$. Then, observe that $g_0 > 0$ for all t and p and use the assumption that $\gamma^2 < \gamma$ which yields $g_0 + g_1\gamma > (g_0 + g_1)\gamma$. Further, we have $g_0 + \ldots + g_i \geq 0$, for all $1 \leq i \leq 3$, and for all $t \geq p \geq 3$, which implies that $g(\gamma) > (g_0 + \ldots + g_3)\gamma^3 \geq 0$ for all $t \geq p \geq 3$. ✓

(2) The minimum of the h_1 parabola is located at $x_{min} = t/(tp - t - 1)$. The value of x_1 is negative and the value of x_2 is positive. Calculating the Manhattan distance (L^1 norm), $d(x_{min}, x_1) = x_{min} + |x_1|$, while $d(x_{min}, x_2) = x_{min} - x_2 < x_{min} + |x_1|$ if $x_2 < x_{min}$, or $d(x_{min}, x_2) = x_2 - x_{min} \overset{|x_2| \leq |x_1|}{<} x_{min} + |x_1|$ if $x_2 > x_{min}$. Hence, $d(x_{min}, x_1) > d(x_{min}, x_2)$. Since h_1 is a quadratic function, it follows that $h_1(x_2) < h_1(x_1)$. ✓

(3) The proof of $x_2 < 1$ is given in Appendix C, Lemma 17.

□

A third equivalence class, $l = k$, is introduced by the representative sequence $[1, 2, 2, 3, 3, \ldots, \frac{p+1}{2}, \frac{p+1}{2}]$ for odd p, or $[1, 2, 3, 3, \ldots, \frac{p+2}{2}, \frac{p+2}{2}]$ for even p. Dependent on the number of periods p, the properties of h_k differ as follows:

For odd p, the sequence class function h_k of equation (2.7) is given as:

$$h_k(x) = \frac{(p-1)(p-1+2\gamma)}{(\gamma+1)(\gamma+p)} + \frac{(p-3)(\gamma+p-2)}{(\gamma+1)(\gamma+p)} \cdot x$$
$$+ \frac{(p-1)\gamma^2 + (3p-5)t\gamma + (p^2 - 3p + 4)t - p + 1}{t(1-\gamma^2)(\gamma+p)} \cdot x^2,$$

for which $R_k = 1 + (p-1)/(\gamma+1)$. The functions h_1 and h_k intersect twice at

$$x_{-k,k} = \frac{\mp\sqrt{W} - V}{Z}, \qquad (3.14)$$

in the interval $(0, 1)$ of γ, in which

3 Deriving General Formulas and Some of Their Properties

$$W = t(1-\gamma)\{[4(p-1)^2(p^2-p-1)t - 4p(p-1)^2]\gamma^5 + [4(p-1)(2p^4-5p^3+p^2+3p+1)t - 4p(p-1)^3]\gamma^4 + [(2p^2-3p-1)(2p^4-5p^3+5p^2+4p-4)t]\gamma^3 + [(6p^5-7p^4-16p^3+17p^2+8p+4)t + 4p(p-1)^3]\gamma^2 + [p(-p^5+8p^4-13p^3+14p^2-8p-12)t + 4p(p-1)^2]\gamma + (p^4-6p^3+13p^2-8p+4)p^2t\},$$

$$V = t(1-\gamma)[2(p-1)\gamma^2 + (3p^2-3p-2)\gamma + (p^2-3p+4)p]$$

and

$$Z = 2\gamma[((p-1)^2t-p+1)\gamma^3 + p(p-1)^2t\gamma^2 + ((2p^2-3p-1)t+p-1)\gamma - p(p-3)t].$$

For even p, the sequence class function h_k of equation (2.7) is given as

$$h_k(x) = \frac{(p-1)(p-1+2\gamma)}{(\gamma+1)(\gamma+p)} + \frac{(p-3)(\gamma+p-2)}{(\gamma+1)(\gamma+p)} \cdot x$$
$$+ \frac{((p-1)\gamma^2 + (3p-5)t\gamma + (p^2-3p+4)t - p+1)}{t(1-\gamma^2)(\gamma+p)} \cdot x^2,$$

for which $R_k = 2 + (p-2)/(\gamma+1)$. Similar to equation (3.14), in which $\gamma \in (0,1)$, the functions h_1 and h_k intersect twice at

$$x_{-k,k} = \frac{\mp\sqrt{W} - V}{Z}, \qquad (3.15)$$

in which

$$W = t(1-\gamma)\{[4(p-2)^2(4p^2-2p-1)t - 8(p-2)^2p]\gamma^5 + [4(p-2)(4p^4-13p^3+7p^2+7p+2)t - 4(p-2)^2(p-1)p]\gamma^4 + [4(p^2-p-1)(p^4-5p^3+10p^2-8p-4)t + 4(p-2)^2p]\gamma^3 + [4(p^5-p^4-11p^3+14p^2+12p+4)t + 4(p-2)^2(p-1)p]\gamma^2 + [(-p^5+12p^4-40p^3+60p^2-36p-32)pt + 4(p-2)^2p]\gamma + (p^4-8p^3+24p^2-24p+4)p^2t\},$$

$$V = t(1-\gamma)[2(p-2)\gamma^2 + 4(p^2-2p-1)\gamma + p(p^2-4p+6)]$$

and

$$Z = 2\gamma[((p-2)(2p-1)t-p+2)\gamma^3 + (p-2)(p-1)pt\gamma^2 + ((2p^2-5p-2)t+p-2)\gamma - (p-4)pt].$$

Proposition 4. Let $\gamma \in (0.3, 1)$ and $t \geq p \geq 5$. The x-coordinate of the minimum of the intersections of sequence class functions h_1 and h_k is $x_k \in (0,1)$ of equation (3.14) for odd p, or $x_k \in (0,1)$ of equation (3.15) for even p, respectively.

Proof. Follows from Lemmas 18 and 19 of Appendix C. □

4 Optimal Designs

This chapter introduces optimal designs for sequences with $p = 3, 4, 5$ and 6 periods. The equivalence classes representation given in the tables take the following rules into account: All equivalence classes are grouped according to their R_u. The sum R_u of all row sums of S_{du}^{-1}, and $c_{11}(l)$ simply depend on the number of appearances of each treatment $j = 1, \ldots, t$ in the sequence. The order of the treatments in the sequence can be neglected. Thus, several equivalence classes yield the same R_u, and $c_{11}(l)$ respectively, but their sequence class functions h_l might be different. The groups of equivalent $c_{11}(l)$ are arranged in descending order. Within each group, the equivalence classes are listed in descending order of $c_{12}(l)$. The arrangement is determined by $c_{22}(l)$ for identical values of $c_{11}(l)$ and $c_{12}(l)$. All classes $l = 1, \ldots, K$ are labelled according to their generation by replacing each treatment of a period repeated times. The starting sequence is class $[1, 2, \ldots, p]$. Equivalent sequences are eliminated, following Definition 2. However, there are several equivalence classes with identical h_l functions among the K listed classes. This is because of the assumed model of section 2.1. Therefore, some values of $\{1, \ldots, K\}$ do not appear in the class listings. The representative sequences of identical h_l functions are listed in one group of the corresponding h_l.

Several calculations and term transformations of this thesis have been performed using the computer algebra wxMaxima 0.7.1, which is implemented on Maxima 5.11.0.

4.1 Sequence Length $p = 3$

Restricting the sequence length to $p = 3$ periods, there are 5 equivalence classes, which are listed in Table 4.1.

Referring to section 3.4, the intersection point x_2 of $h_1(x) = h_2(x)$ is located at

$$x_2 = \frac{\sqrt{t(1-\gamma)[(5t-3)\gamma^3 + 13t\gamma^2 + (9t+3)\gamma + 9t]} - t(1-\gamma)(\gamma + 3)}{(2t-1)\gamma^2 + (4t+1)\gamma}. \quad (4.1)$$

The parameter domains are $\gamma \in (0, 1)$ and $t \geq 3$. Define $RT_{x_2} = t(1-\gamma)[(5t-3)\gamma^3 + 13t\gamma^2 + (9t+3)\gamma + 9t]$; it will be referred to as the root term of x_2.

4 Optimal Designs

$l : [...]$	$h_l =$	$c_{11}(l)$	$+2c_{12}(l)x$	$+c_{22}(l)x^2$
$R_u = 3$				
$1 : [123]$	$h_1 =$	2	$-\frac{4}{3} \cdot x$	$+\frac{2(2t-1)}{3t} \cdot x^2$
$R_u = (\gamma+3)/(\gamma+1)$				
$2 : [122]$	$h_2 =$	$\frac{4}{(3+\gamma)}$	$+0$	$+\frac{2(\gamma+2t-1)}{t(1-\gamma)(3+\gamma)} \cdot x^2$
$4 : [112]$	$h_4 =$	$\frac{4}{(3+\gamma)}$	$-\frac{2}{(3+\gamma)} \cdot x$	$+\frac{2(t-1)}{t(1-\gamma)(3+\gamma)} \cdot x^2$
$3 : [121]$	$h_3 =$	$\frac{4}{(3+\gamma)}$	$-\frac{6}{(3+\gamma)} \cdot x$	$+\frac{2(t(2-\gamma)-1)}{t(1-\gamma)(3+\gamma)} \cdot x^2$
$R_u = 3/(2\gamma+1)$				
$5 : [111]$	$h_5 =$	0	$+0$	$+\frac{2(t-1)}{3t(1-\gamma)} \cdot x^2$

Table 4.1: Equivalence classes l, their representative sequences [...] and h_l functions for sequence length $p = 3$.

Lemma 3. For any $t \geq p = 3$, $\gamma \in (0,1)$, and x_2 being the intersection point of equivalence class functions $h_1(x) = h_2(x)$, cf. equation (4.1), it follows that $h_2(x_2) = \max_{l \in \{2,...,5\}} h_l(x_2)$.

Proof. Look at the equivalence class functions h_l of Table 4.1:

a) $h_2(x) > h_3(x)$ and $h_2(x) > h_4(x)$ for all $x > 0$, because $c_{11}(2) = c_{11}(3) = c_{11}(4)$; $c_{12}(2) > c_{12}(3)$ and $c_{22}(2) > c_{22}(3)$; $c_{12}(2) > c_{12}(4)$ and $c_{22}(2) > c_{22}(4)$ for all $\gamma \in (0,1)$ and $t \geq 3$. The results are reproducible by considering Lemmas 1 and 2 for the comparison of $c_{11}(l)$ and $c_{12}(l)$, respectively.

b) $h_2(x) > h_5(x)$ for all $x > 0$, because $c_{ij}(2) > c_{ij}(5)$, $1 \leq i, j \leq 2$, $\gamma \in (0,1)$ and $t \geq 3$.

As $x_2 > 0$, Lemma 3 follows from a) and b). $\square$

Lemma 4. For any $t \geq 16$ and $p = 3$, the parameters $0 < \gamma_1 < \gamma_2 < 1$ are given by

$$\gamma_{1/2} = \frac{\mp\sqrt{(4t-3)(4t^3 - 67t^2 + 52t - 12)} + (4t-5)t}{2(2t-1)(5t-3)}.$$

Assume $\gamma \in (\gamma_1, \gamma_2)$ exists, and observe that $x_{min} = t/(2t-1)$ is the abscissa of the minimum of equivalence class function $h_1(x)$, then, $h_1(x_{min}) = \max_{l \in \{1,...,5\}} h_l(x_{min})$.

Proof. As $x_{min} > 0$ and $h_2(x) > h_l(x)$ for all $l \in \{3,4,5\}$ and all $x > 0$, it is sufficient to verify whether $h_1(x_{min}) > h_2(x_{min})$. Thus, calculate the expression

$$(h_2 - h_1)(x_{min}) = \frac{2[(2t-1)(5t-3)\gamma^2 - t(4t-5)\gamma + 3(2t-1)]}{3(2t-1)^2(1-\gamma)(\gamma+3)}.$$

The solution of the quadratic equation $(2t-1)(5t-3)\gamma^2 - t(4t-5)\gamma + 3(2t-1) = 0$ for all $\gamma \in (0,1)$ and all $t \geq 3$ implies

$$(h_2 - h_1)(x_{min}) \begin{cases} < 0 & \Leftrightarrow \gamma \in (\gamma_1, \gamma_2) \\ \geq 0 & \Leftrightarrow \gamma \notin (\gamma_1, \gamma_2) \end{cases}.$$

Hence, $h_1(x_{min}) > h_2(x_{min})$ for all $\gamma \in (\gamma_1, \gamma_2)$, and Lemma 4 follows. $\square$

Theorem 1. For any $t \geq p = 3$ and $\gamma \in (0,1)$, the proportion $\alpha(\gamma) \in [0,1]$ is given by

$$\alpha(\gamma) = \frac{(1-\gamma)(\gamma+3)\left[3t(\gamma+2t-1) - (2t-1)\sqrt{RT_{x_2}}\right]}{\gamma[(2t-1)\gamma + 4t + 1]\sqrt{RT_{x_2}}}.$$

If, additionally, $t \geq 16$, the parameters $0 < \gamma_1 < \gamma_2 < 1$ are given as in Lemma 4.
The optimal results are as follows:
For all $t < 16$, or if $t \geq 16$ and $\gamma \notin (\gamma_1, \gamma_2)$, an approximate design d^* is optimal iff $(1 - \alpha(\gamma)) \cdot 100\%$ of its sequences are selected from class 1 with representative sequence $[1,2,3]$ and $\alpha(\gamma) \cdot 100\%$ of its sequences from class 2 with representative sequence $[1,2,2]$. For $t \geq 16$ and $\gamma \in (\gamma_1, \gamma_2)$, all sequences of d^* are representatives of equivalence class 1.

Proof. Theorem 1 claims that x_{d^*} of Proposition 2 in which the $\min_x \max_l h_l(x)$ is being realized, is either x_2 or x_{min}, the x-coordinate of the minimum of h_1. In this manner, four points need to be verified:

1. For all $t < 16$ and $\gamma \in (0,1)$ or $t \geq 16$ and $\gamma \notin (\gamma_1, \gamma_2)$: $h_2(x_2) > h_l(x_2)$ for all $3 \leq l \leq 5$.

2. For all $t < 16$ and $\gamma \in (0,1)$ or $t \geq 16$ and $\gamma \notin (\gamma_1, \gamma_2)$: $\text{sign } h'_1(x_2) \neq \text{sign } h'_2(x_2)$.

3. For all $t \geq 16$ and $\gamma \in (\gamma_1, \gamma_2)$: $h_1(x_{min}) > h_l(x_{min})$ for all $2 \leq l \leq 5$.

4. The formula for $\alpha(\gamma)$.

Properties 1 and 3 are proved by Lemmas 3 and 4, respectively. ✓
As demanded in point 4, the proportion of sequences $\alpha(\gamma) \in (0,1)$ of equivalence class 2 needs to be determined. To this end, use equation (2.8) and put $\alpha = \alpha(\gamma, x)$, then $q_{d^*}(x) = $

4 Optimal Designs

$\alpha n h_2(x) + (1-\alpha)n h_1(x)$. The formula of $\alpha(\gamma, x)$ is derived by setting

$$0 \stackrel{!}{=} \frac{\delta q_{d^*}}{\delta x} = \frac{4}{3+\gamma}\alpha\left(\frac{2t-1+\gamma}{t(1-\gamma)}x\right) - \frac{4}{3}(1-\alpha) + 4(1-\alpha)\frac{(2t-1)}{3t}x$$

$$\Leftrightarrow \alpha(\gamma, x) = \frac{[(1-2t)x + t](1-\gamma)(\gamma + 3)}{[(2t-1)\gamma + 4t + 1]\gamma x + t(\gamma + 3)(1-\gamma)}.$$

Substitution of $x = x_2$ in the formula of $\alpha(\gamma, x)$ provides $\alpha(\gamma)$. A graphical presentation of $\alpha(\gamma)$ for different t is displayed in Figure 4.1. ✓

Referring to statement 2, there is one major condition in order to achieve that $\alpha \in (0,1)$ by using equation $0 \stackrel{!}{=} \alpha n h_2'(x) + (1-\alpha)n h_1'(x)$ to derive proportion $\alpha(\gamma)$. The condition is given as sign $h_1'(x_2) \neq$ sign $h_2'(x_2)$. The proportion $\alpha(\gamma) < 0$ or $\alpha(\gamma) > 1$ iff sign $h_1'(x_2) =$ sign $h_2'(x_2)$. This is not valid for an equivalence class proportion. Thus, to prove 2, it is sufficient to analyze if $a(\gamma)$, or $\alpha(\gamma, x)$ respectively, is nonnegative and less or equal to 1 in the described domains of parameters t and γ. For this, observe

$$\alpha(\gamma, x = x_2) \stackrel{!}{=} 0 \Leftrightarrow [(1-2t)x_2 + t] = 0$$
$$\stackrel{(4.1)}{\Leftrightarrow} t\gamma[(2t-1)\gamma + 4t + 1][(2t-1)(5t-3)\gamma^2 - t(4t-5)\gamma + 3(2t-1)] = 0$$
$$\Leftrightarrow \gamma \in \{\gamma_1, \gamma_2, -(4t+1)/(2t-1), 0\}$$

The last two elements, $-(4t+1)/(2t-1)$, and 0, are not in the domain $(0,1)$ of γ. The boundaries $\gamma_{1/2}$ are defined for all $t \geq 16$. The proportion $\alpha(\gamma) \geq 0$, iff $\gamma \notin (\gamma_1, \gamma_2)$, since $t\gamma[(2t-1)\gamma + 4t + 1] > 0$ for all $\gamma \in (0,1)$ and $t > 0$, and $[(2t-1)(5t-3)\gamma^2 - t(4t-5)\gamma + 3(2t-1)]$ is a convex parabola of γ.

In order to specify function $\alpha(\gamma)$ as a proportion, we have to examine whether $\alpha(\gamma) \leq 1$. For this purpose, rewrite the numerator of $\alpha(\gamma, x)$ as $t(1-\gamma)(\gamma+3) - \underbrace{(2t-1)(1-\gamma)(\gamma+3)}_{>0}x$ and the denominator of $\alpha(\gamma, x)$ as $t(1-\gamma)(\gamma+3) + \underbrace{[(2t-1)\gamma + 4t + 1]}_{>0}\gamma x$. It is easy to derive that $\alpha(\gamma, x) < 1$ for all $x, t > 0$ and $\gamma \in (0,1)$ because the numerator of $\alpha(\gamma, x)$ must be smaller than its denominator. The value $x = x_2$ is positive and it follows that $\alpha(\gamma) \leq 1$ for all $t \geq 3$ and all $\gamma \in (0,1)$.

The statement sign $h_1'(x_2) \neq$ sign $h_2'(x_2)$ holds for all $t < 16$ and $\gamma \in (0,1)$ or $t \geq 16$ and $\gamma \notin (\gamma_1, \gamma_2)$, since $0 \leq \alpha(\gamma) \leq 1$. ✓

Theorem 1 follows from properties 1 through 4. □

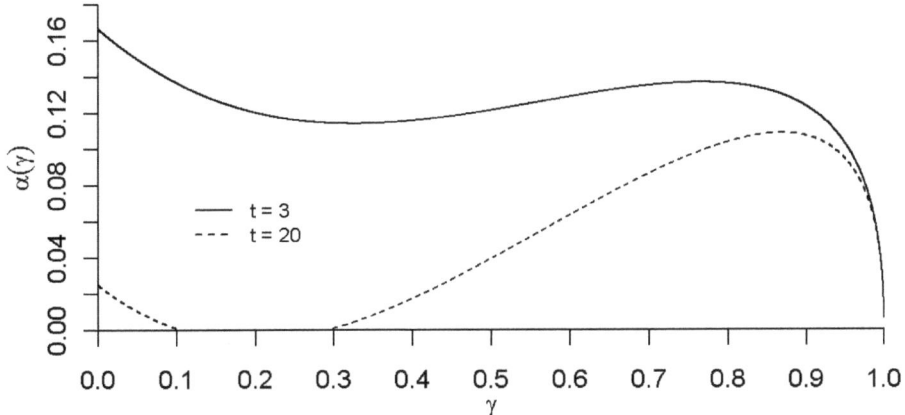

Figure 4.1: Proportions $\alpha(\gamma)$ of equivalence class 2 sequences for an approximate optimal design with $p = 3$ periods.

4.2 Sequence Length $p = 4$

In the case of 4 periods for each sequence, there are 14 different equivalence classes for $l = 1, \ldots, 15$. All classes are listed in Table 4.2.

The intersection point x_2 of section 3.4, in which $h_1(x) = h_2(x)$, is located at

$$x_2 = \frac{\sqrt{t(1-\gamma)[(55t-8)\gamma^3 + (57t+4)\gamma^2 + (4-4t)\gamma + 36t]} - t(1-\gamma)(\gamma+6)}{(7t-1)\gamma^2 + (5t+1)\gamma}. \qquad (4.2)$$

The parameter domains are $\gamma \in (0,1)$ and $t \geq 4$. Let $RT_{x_2} = t(1-\gamma)[(55t-8)\gamma^3 + (57t+4)\gamma^2 + (4-4t)\gamma + 36t]$ be the root term of x_2 in the case of $p = 4$.

The equivalence classes $2, \ldots, 14$ are divided into groups of identical R_u. The equivalence class l belongs to

group A $\Leftrightarrow l \in L_A = \{2, 3, 4, 5, 8, 11\}$;

group B $\Leftrightarrow l \in L_B = \{7, 9, 12\}$;

group C $\Leftrightarrow l \in L_C = \{6, 10, 14\}$.

The intention is to determine the maximal h_l functions for each defined group, assuming $x > 0$. The coefficients $c_{11}(l)$ within each group are identical.

A) If $l \in L_A$, $h_2(x)$ is maximal among the h_l. This result follows from $c_{12}(2) > c_{12}(l)$ and $c_{22}(2) \geq c_{22}(l)$.

4 Optimal Designs

$l : [...]$	$h_l =$	$c_{11}(l)$	$+2c_{12}(l)x$	$+c_{22}(l)x^2$
$R_u = 4$				
1 : [1234]	$h_1 =$	3	$-\frac{3}{2} \cdot x$	$+\frac{3(3t-1)}{4t} \cdot x^2$
Group A: $R_A = 2(\gamma+2)/(\gamma+1)$				
2 : [1233]	$h_2 =$	$\frac{(5+\gamma)}{(2+\gamma)}$	$-\frac{\gamma}{(2+\gamma)} \cdot x$	$+\frac{(9t-3-2(t-1)\gamma-(t-1)\gamma^2)}{2t(\gamma+2)(1-\gamma)} \cdot x^2$
5 : [1223]	$h_5 =$	$\frac{(5+\gamma)}{(2+\gamma)}$	$-\frac{1}{(2+\gamma)} \cdot x$	$+\frac{(7t-3+2\gamma-(t-1)\gamma^2)}{2t(\gamma+2)(1-\gamma)} \cdot x^2$
11 : [1123]	$h_{11} =$	$\frac{(5+\gamma)}{(2+\gamma)}$	$-\frac{(1+\gamma)}{(2+\gamma)} \cdot x$	$+\frac{(7t-3-2t\gamma-2\gamma^2)}{2t(\gamma+2)(1-\gamma)} \cdot x^2$
3 : [1232]	$h_3 =$	$\frac{(5+\gamma)}{(2+\gamma)}$	$-\frac{4}{(2+\gamma)} \cdot x$	$+\frac{(9t-3-2(t-1)\gamma-(t-1)\gamma^2)}{2t(\gamma+2)(1-\gamma)} \cdot x^2$
4 : [1231]	$h_4 =$	$\frac{(5+\gamma)}{(2+\gamma)}$	$-\frac{(4+\gamma)}{(2+\gamma)} \cdot x$	$+\frac{(9t-3-4t\gamma-2t\gamma^2)}{2t(\gamma+2)(1-\gamma)} \cdot x^2$
8 : [1213]	$h_8 =$	$\frac{(5+\gamma)}{(2+\gamma)}$	$-\frac{5}{(2+\gamma)} \cdot x$	$+\frac{(7t-3-4t\gamma)}{2t(\gamma+2)(1-\gamma)} \cdot x^2$
Group B: $R_B = 4/(\gamma+1)$				
12 : [1122]	$h_{12} =$	$\frac{2}{(1+\gamma)}$	$+\frac{1}{(1+\gamma)} \cdot x$	$+\frac{(7t-3+(5t-1)\gamma)}{4t(1-\gamma^2)} \cdot x^2$
7 : [1221]	$h_7 =$	$\frac{2}{(1+\gamma)}$	$-\frac{1}{(1+\gamma)} \cdot x$	$+\frac{(7t-3+(5t-1)\gamma)}{4t(1-\gamma^2)} \cdot x^2$
9 : [1212]	$h_9 =$	$\frac{2}{(1+\gamma)}$	$-\frac{3}{(1+\gamma)} \cdot x$	$+\frac{(7t-3-(3t+1)\gamma)}{4t(1-\gamma^2)} \cdot x^2$
Group C: $R_C = 2(\gamma+2)/(2\gamma+1)$				
6 : [1222]	$h_6 =$	$\frac{4(1-\gamma)}{3}$	$+\frac{4(1-\gamma)}{9} \cdot x$	$+\frac{2(17t-9-(4t-18)\gamma+(5t-9)\gamma^2)}{27t(1-\gamma)} \cdot x^2$
14 : [1112]	$h_{14} =$	$\frac{4(1-\gamma)}{3}$	$-\frac{4(1-\gamma)}{9} \cdot x$	$+\frac{2(7t-7+(t-1)\gamma+(t-1)\gamma^2)}{27t(1-\gamma)} \cdot x^2$
10 : [1211] 13 : [1121]	$h_{10} =$	$\frac{4(1-\gamma)}{3}$	$-\frac{4(1-\gamma)}{3} \cdot x$	$+\frac{2(17t-7-(4t+1)\gamma+(5t-1)\gamma^2)}{27t(1-\gamma)} \cdot x^2$
$R_D = 5/(4\gamma+1)$				
15 : [1111]	$h_{15} =$	0	$+0$	$+\frac{3(t-1)}{4t(1-\gamma)} \cdot x^2$

Table 4.2: Equivalence classes l, their representative sequences [...] and h_l functions for sequence length $p = 4$.

4.2 Sequence Length $p = 4$

B) Examine group B, the maximal h_l function is $h_{12}(x)$ because $c_{12}(12) > c_{12}(l)$ and $c_{22}(12) \geq c_{22}(l)$ for all $l \in L_B$.

C) The maximal h_l of group C connected with h_{15} turns out to be either h_6 or h_{10}. This follows from $c_{11}(6) + 2c_{12}(6)x > c_{11}(l) + 2c_{12}(l)x$ for all $l \in L_C \cup \{15\}$. Furthermore, $c_{22}(6)$ or $c_{22}(10)$ is maximal for the $c_{22}(l)$ of $l \in L_C \cup \{15\}$. Dependent on the correlation γ, the value of $c_{22}(10)$ increases rapidly, such that its h_l even exceeds h_6; for all other $l \in L_C \cup \{15\} \setminus \{10\}$, we have $c_{22}(6) \geq c_{22}(l)$.

D) Unite groups B and $C \cup \{15\}$. Then, $h_{12}(x) = \max_{l \in \{6,10,12\}} h_l(x)$. Observe that

$$c_{11}(12) + 2c_{12}(12)x - c_{11}(l) - 2c_{12}(l)x \geq \frac{4(x+3)\gamma^2 + 5x + 6}{9(\gamma+1)} > 0, \quad \text{for } l = 6, 10.$$

Moreover,

$$c_{22}(12) - c_{22}(6) = \frac{(72 - 40t)\gamma^3 - (8t + 72)\gamma^2 + (31t - 99)\gamma + 53t - 9}{108t(1 - \gamma)(\gamma + 1)} > 0$$

as well as

$$c_{22}(12) - c_{22}(10) = \frac{(8 - 40t)\gamma^3 + (16 - 8t)\gamma^2 + (31t + 37)\gamma + 53t - 25}{108t(1 - \gamma)(\gamma + 1)} > 0.$$

The positive values can be confirmed by defining both numerators as functions $G_l(\gamma) = g_0 + g_1\gamma + g_2\gamma^2 + g_3\gamma^3$ with

$G_6(\gamma) = (72 - 40t)\gamma^3 - (8t + 72)\gamma^2 + (31t - 99)\gamma + 53t - 9$

and

$G_{10}(\gamma) = (8 - 40t)\gamma^3 + (16 - 8t)\gamma^2 + (31t + 37)\gamma + 53t - 25$.

Since $t \geq 4$, the coefficients g_0 and g_1 are positive for both functions $G_6(\gamma)$ and $G_{10}(\gamma)$. As $g_0 + g_1 + \ldots + g_i > 0$, for $i = 2, 3$, $\gamma \in (0, 1)$, and all $t \geq 4$, we have that $G_l(\gamma) > (g_0 + \ldots + g_3)\gamma^3 > 0$ for $l = 6, 10$ and all $t \geq 4$. The denominators of both ratios are identical and positive for $\gamma \in (0, 1)$. Thus, $c_{22}(12) > c_{22}(l)$ for $l = 6, 10$ and all $t \geq 4$.

Lemma 5. For any $t \geq p = 4$, $\gamma \in (0, 1)$, and x_2 being the intersection point of equivalence class functions $h_1(x) = h_2(x)$, cf. equation (4.2), it follows that $h_2(x_2) = \max_{l \in \{2,\ldots,15\}} h_l(x_2)$.

Proof. Refer to the equivalence class functions h_l of Table 4.2 and use the results of A) through D) above. There are just two equivalence class functions, h_2 and h_{12}, which exceed all other

4 Optimal Designs

h_l functions, $l \in \{3, \ldots, 11, 13, 14, 15\}$, for $x > 0$. Since $x_2 > 0$, $h_2(x_2) = \max_{l \in \{2,\ldots,15\}} h_l(x_2)$ iff $h_2(x_2) \geq h_{12}(x_2)$. Accordingly, analyze

$$(h_2 - h_{12})(x) = \frac{\gamma^2 + 4\gamma + 1}{(\gamma+1)(\gamma+2)} - \frac{\gamma^2 + 2\gamma + 2}{(\gamma+1)(\gamma+2)} x$$
$$- \frac{2(t-1)\gamma^3 + (11t-7)\gamma^2 + 3(t-1)\gamma - 4t}{4t(1-\gamma)(\gamma+1)(\gamma+2)} x^2.$$

Set $x = x_2$, and $h_2 - h_{12}$ turns out to be

$$(h_2 - h_{12})(x_2) = \frac{a(\gamma) + b(\gamma)\sqrt{RT_{x_2}}}{2\gamma^2(\gamma+1)[(7t-1)\gamma + 5t + 1]^2}.$$

The expression $a(\gamma) := a_0 + a_1\gamma + \ldots + a_5\gamma^5$ is given as

$a(\gamma) = 6(5t^2 + 6t - 1)\gamma^5 + (21t^2 + 111t - 8)\gamma^4 + 2(103t^2 + 5)\gamma^3 + (17t^2 + 71t + 4)\gamma^2 - 2t(29t - 35)\gamma + 72t^2.$

The expression $b(\gamma) := b_0 + b_1\gamma + b_2\gamma^2 + b_3\gamma^3$ is a substitute for

$b(\gamma) = -12t\gamma^3 + (9t - 17)\gamma^2 + (3t - 11)\gamma - 12t.$

Again, use $a_0 > 0$ and $a_0 + \ldots + a_i > 0$ for $1 \leq i \leq 5$. It can be concluded that $a(\gamma) \geq (a_0 + a_1 + \ldots + a_5)\gamma^5 > 0$.

Apply the same method to $-b(\gamma)$ to obtain $b(\gamma) < 0$.

The denominator of $(h_2 - h_{12})(x_2)$ is positive. Hence, in order to determine wether $(h_2 - h_{12})(x_2)$ is positive, it is necessary to confirm that $a(\gamma) + b(\gamma)\sqrt{RT_{x_2}} > 0$. Thus, subtract $b(\gamma)\sqrt{RT_{x_2}}$ on both sides of the inequality and square them afterwards to obtain

$$a^2(\gamma) - b^2(\gamma)RT_{x_2} > 0$$
$$\Leftrightarrow 4\gamma^2[(7t-1)\gamma + 5t + 1]^2 \cdot g(\gamma) > 0$$
$$\Leftrightarrow g(\gamma) := g_0 + g_1\gamma + \ldots + g_6\gamma^6 > 0,$$

in which

$g(\gamma) = 9(5t^2 + 2t + 1)\gamma^6 - 3(39t^2 - 37t - 14)\gamma^5 + (165t^2 - 42t + 61)\gamma^4 + 2(33t^2 + 33t + 14)\gamma^3 - 4(26t^2 - 29t - 1)\gamma^2 + t(73t + 19)\gamma + 16t^2.$

In order to analyze $g(\gamma)$, observe that g_4 and g_6 are positive for all $t \geq 4$. The coefficient g_5 is negative, but $g_4\gamma^4 + g_5\gamma^5 > (g_4 + g_5)\gamma^5 > 0$, which implies that $g(\gamma)$ is positive iff $\sum_{i=0}^{3} g_i\gamma^i > 0$. Therefore, define $G_{03}(\gamma) := g_0 + \ldots + g_3\gamma^3$. The third derivative of $G_{03}(\gamma)$ is positive because $g_3 > 0$. $G'''_{03}(\gamma) > 0$ implies that $\partial G_{03}(\gamma)/\partial\gamma$, the function of the slope of $G_{03}(\gamma)$, is convex. The curvature of $G_{03}(\gamma)$ is the slope of $G'_{03}(\gamma)$, which is given by the second derivative of

4.2 Sequence Length $p = 4$

$G_{03}(\gamma)$. $G_{03}''(\gamma)$ equals 0 iff $\gamma = \gamma_0 := \frac{2(26t^2-29t-1)}{3(33t^2+33t+14)}$. The value of $G_{03}''(\gamma_0)$ is equal to $\frac{1819t^4+21172t^3-1365t^2+334t-8}{3(33t^2+33t+14)}$ and positive for all $t \geq 4$.

Because $G_{03}'(\gamma)$ is convex, $G_{03}''(\gamma_0)$ must be the minimum of $G_{03}''(\gamma)$. Hence, the slope of $G_{03}(\gamma)$ is nonnegative for all $\gamma \in (0,1)$.

Thus, $G_{03}(\gamma)$ is monotonous and increasing in γ, and the local minima of $G_{03}(\gamma)$ is given at $\gamma \searrow 0$. The minimum $G_{03}(\gamma \searrow 0) = g_0 = 16t^2$ is positive. This leads to $G_{03}(\gamma)$ being positive for all $t \geq 4$ and all $\gamma \in (0,1)$.

It follows that $g(\gamma)$ is positve and the conclusion is: $h_2(x_2) > h_{12}(x_2)$ for all $\gamma \in (0,1)$ and all $t \geq 4$. □

Lemma 6. For any $t \geq p = 4$, the parameters $0 < \gamma_1 < \gamma_2 < 1$ are given as

$$\gamma_{1/2} = \frac{\mp\sqrt{3721t^4 - 7774t^3 + 5137t^2 - 1416t + 144} + 61t^2 - 35t + 4}{2(73t^2 - 47t + 8)}.$$

Assume $\gamma \in (\gamma_1, \gamma_2)$, and observe that $x_{min} = t/(3t-1)$ is the abscissa of the minimum of equivalence class function $h_1(x)$. Then $h_1(x_{min}) = \max_{l \in \{1,\ldots,15\}} h_l(x_{min})$.

Proof. As $x_{min} > 0$, we can use the results of A) through D), i.e., either h_2 or h_{12} is the maximum of all other h_l functions for $l \in \{2,\ldots,15\}$ for any $x > 0$. Thus, iff $h_1(x_{min}) > h_2(x_{min})$ and $h_1(x_{min}) > h_{12}(x_{min})$, $h_1(x_{min}) = \max_{l=1,\ldots,15} h_l(x_{min})$. Proceeding this way, consider the first difference of interest

$$(h_2 - h_1)(x_{min}) = \frac{4(3t-1) - (61t^2 - 35t + 4)\gamma + (73t^2 - 47t + 8)\gamma^2}{4(3t-1)^2(1-\gamma)(\gamma+2)}.$$

The solution of the quadratic equation $4(3t-1) - (61t^2 - 35t + 4)\gamma + (73t^2 - 47t + 8)\gamma^2 = 0$ for all $\gamma \in (0,1)$ and all $t \geq 4$ gives

$$(h_2 - h_1)(x_{min}) \begin{cases} < 0 & \Leftrightarrow \gamma \in (\gamma_1, \gamma_2) \\ \geq 0 & \Leftrightarrow \gamma \notin (\gamma_1, \gamma_2) \end{cases}.$$

Hence, $h_1(x_{min}) > h_2(x_{min})$ for $\gamma \in (\gamma_1, \gamma_2)$ and $\gamma_{1/2}$ being defined as in Lemma 6.

Analyze the expression

$$(h_{12} - h_1)(x_{min}) = \frac{3(3t-1)(11t-4)\gamma^2 - (79t^2 - 51t + 8)\gamma - 2(4t^2 - 7t + 2)}{4(3t-1)^2(1-\gamma^2)}.$$

Solving the quadratic equation $3(3t-1)(11t-4)\gamma^2 - (79t^2 - 51t + 8)\gamma - 2(4t^2 - 7t + 2) = 0$ for all $\gamma \in (0,1)$ and all $t \geq 4$ yields

$$(h_{12} - h_1)(x_{min}) \begin{cases} = 0 & \Leftrightarrow \gamma = \gamma_{5/6} = \frac{\mp\sqrt{9409t^4 - 15810t^3 + 9697t^2 - 2592t + 256} + 79t^2 - 51t + 8}{6(3t-1)(11t-4)} \\ < 0 & \Leftrightarrow \gamma \in (\gamma_5, \gamma_6) \\ > 0 & \Leftrightarrow \gamma \notin (\gamma_5, \gamma_6) \end{cases}.$$

4 Optimal Designs

The parameter γ_5 is negative and it follows immediately that $\gamma_5 < \gamma_1$. Next, consider the parameters γ_2 and γ_6 as functions of t. Both functions $\gamma_2(t)$ and $\gamma_6(t)$ are increasing in t as their slopes are positive for all $t \geq 4$. Observe that $\gamma_2(t \nearrow \infty) = 61/73 < (\sqrt{96369} + 267)/660 = \gamma_6(t \searrow 4)$, i.e., the maximum of the function $\gamma_2(t)$ is less than the minimum of the function $\gamma_6(t)$, $t \geq 4$. Therefore, $\gamma_2 < \gamma_6$ for all $t \geq 4$. Now we have $(\gamma_1, \gamma_2) \subset (\gamma_5, \gamma_6)$. Thus, $h_1(x_{min}) > h_{12}(x_{min})$ for $\gamma \in (\gamma_1, \gamma_2)$. Hence, $h_1(x_{min}) > h_l(x_{min})$ for all $\gamma \in (\gamma_1, \gamma_2)$ and all $l = 2, \ldots, 15$. Lemma 6 follows. $\square$

Theorem 2. For any $t \geq p = 4$ and $\gamma \in (0,1)$, consider the parameters $0 < \gamma_1 < \gamma_2 < 1$ as in Lemma 6. Furthermore, the proportion $\alpha(\gamma) \in [0,1]$ is given by

$$\alpha(\gamma) = \frac{3(1-\gamma)(\gamma+2)[2t(2t\gamma^2 + (3-5t)\gamma + 9t - 3) - (3t-1)\sqrt{RT_{x_2}}]}{\gamma((7t-1)\gamma + 5t + 1)\sqrt{RT_{x_2}}}$$

iff $\gamma \in (0, \gamma_1] \overset{\bullet}{\cup} [\gamma_2, 1)$ and $\alpha(\gamma) := 0$ iff $\gamma \in (\gamma_1, \gamma_2)$.

The optimality results are as follows:

For all $\gamma \in [0, \gamma_1] \overset{\bullet}{\cup} [\gamma_2, 1]$, an approximate design d^* is optimal iff $(1 - \alpha(\gamma)) \cdot 100\%$ of its sequences are selected from class 1 with representative sequence $[1, 2, 3, 4]$ and $\alpha(\gamma) \cdot 100\%$ of its sequences from class 2 with representative sequence $[1, 2, 3, 3]$. Iff $\gamma \in (\gamma_1, \gamma_2)$, all sequences of d^* are representatives of equivalence class 1.

Proof. Theorem 2 indicates that x_{d^*} of Proposition 2 in which $\min_x \max_l h_l(x)$ is being realized, is either x_2 or x_{min}, the x-coordinate of the minimum of h_1. In order to verify this statement, four points need to be treated:

1. For all $\gamma \notin (\gamma_1, \gamma_2)$: $h_2(x_2) > h_l(x_2)$ for all $3 \leq l \leq 15$.

2. For all $\gamma \notin (\gamma_1, \gamma_2)$: sign $h'_1(x_2) \neq$ sign $h'_2(x_2)$.

3. For all $\gamma \in (\gamma_1, \gamma_2)$: $h_1(x_{min}) > h_l(x_{min})$ for all $2 \leq l \leq 15$.

4. The formula of $\alpha(\gamma)$.

Properties 1 and 3 are proved by Lemmas 5 and 6, respectively. ✓

As demanded in statement 4, the proportion $\alpha(\gamma) \in (0,1)$ of equivalence class 2 sequences needs to be determined. To achieve this, use equation (2.8) and set $\alpha = \alpha(\gamma, x)$. We get $q_{d^*}(x) = \alpha n h_2(x) + (1-\alpha) n h_1(x)$. The proportion $\alpha(\gamma)$ is derived by substituting $x = x_2$ in

38

4.2 Sequence Length $p = 4$

the formula of $\alpha(\gamma, x)$ which results from

$$0 \stackrel{!}{=} \frac{\partial q_{d^*}}{\partial x} = \alpha\left(\frac{3}{2} - \frac{\gamma}{2+\gamma} + \frac{3(3t-1) - 2(t-1)\gamma - (t-1)\gamma^2}{t(\gamma+2)(1-\gamma)}x - \frac{3(3t-1)}{2t}x\right)$$
$$- \frac{3}{2} + \frac{3(3t-1)}{2t}x$$
$$\Leftrightarrow \alpha(\gamma, x) = \frac{3[(1-3t)x + t](1-\gamma)(\gamma+2)}{t(1-\gamma)(\gamma+6) + [(7t-1)\gamma + 5t+1]\gamma x}.$$

A graphical presentation of $\alpha(\gamma)$ for different t is displayed in Figure 4.2.   ✓

As in case $p = 3$, the condition $\operatorname{sign} h_1'(x_2) \neq \operatorname{sign} h_2'(x_2)$ needs to be fulfilled. Proportion $\alpha(\gamma)$ is negative or exceeds 1 iff $\operatorname{sign} h_1'(x_2) = \operatorname{sign} h_2'(x_2)$. This, however, is not valid for an equivalence class proportion. To prove 2, it is sufficient to analyze if $a(\gamma)$, or $\alpha(\gamma, x)$ respectively, is nonnegative and less than 1 in the related domains of the parameters t and γ. For this purpose, observe

$$\alpha(\gamma, x = x_2) \stackrel{!}{=} 0 \Leftrightarrow [(1-3t)x_2 + t] = 0$$
$$\stackrel{(4.2)}{\Leftrightarrow} t\gamma[(7t-1)\gamma + 5t + 1][(73t^2 - 47t + 8)\gamma^2 - (61t^2 - 35t + 4)\gamma + 4(3t-1)] = 0$$
$$\Leftrightarrow \gamma \in \{\gamma_1, \gamma_2, -(5t+1)/(7t-1), 0\}.$$

The last two elements, $-(5t+1)/(7t-1)$, and 0, are not in the domain of $\gamma \in (0, 1)$. Since $t\gamma[(7t-1)\gamma + 5t + 1] > 0$ for all $\gamma > 0$ and $t > 0$, and $[(73t^2 - 47t + 8)\gamma^2 - (61t^2 - 35t + 4)\gamma + 4(3t-1)]$ is a convex parabola in γ, proportion $\alpha(\gamma) \geq 0$ iff $\gamma \notin (\gamma_1, \gamma_2)$.

It remains to verify whether $\alpha(\gamma) \leq 1$. Rewrite the numerator of $\alpha(\gamma, x)$ as $t(1-\gamma)(\gamma+6) + 2\gamma t(1-\gamma) - 3(3t-1)(1-\gamma)(\gamma+2)x$. We get $\alpha(\gamma, x) \leq 1$ iff

$$2\gamma t(1-\gamma) - 3(3t-1)(1-\gamma)(\gamma+2)x \leq [(7t-1)\gamma + 5t+1]\gamma x$$
$$\Leftrightarrow -(t+x-tx)\gamma^2 + (t-2x+2tx)\gamma - 3(3t-1)x \leq 0.$$

Substituting $x = x_2 > 0$ and taking into account that $\gamma \in (0, 1)$, we get $-(t+x-tx)\gamma^2 + (t-2x+2tx)\gamma - 3(3t-1)x < [-(t+x-tx) + (t-2x+2tx) - 3(3t-1)x]\gamma = -6tx\gamma < 0$. It follows that $\alpha(\gamma) \leq 1$ for all $\gamma \in (0, 1)$ and $t \geq 4$.

Since $0 \leq a(\gamma) \leq 1$, $\operatorname{sign} h_1'(x_2) \neq \operatorname{sign} h_2'(x_2)$ for all $\gamma \notin (\gamma_1, \gamma_2)$.   ✓

Combining points 1 through 4 provides Theorem 2.   □

4 Optimal Designs

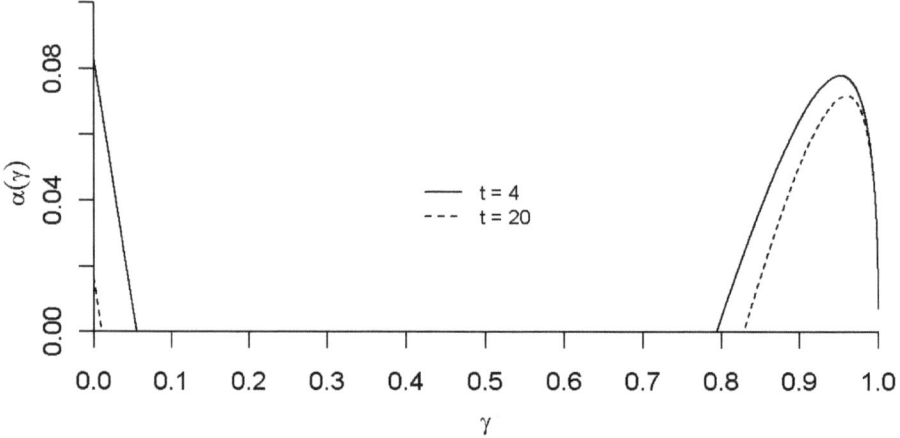

Figure 4.2: Proportions $\alpha(\gamma)$ of equivalence class 2 sequences for an approximate optimal design with $p = 4$ periods.

4.3 Sequence Length $p = 5$

In the case of 5 periods for each sequence, there exist 38 different equivalence classes for $l = 1, \ldots, 52$. The value of the outlined class k of section 3.4 is 19. Nine relevant equivalence classes are listed in Table 4.3. For completeness, the set of all equivalence classes is given in Table B.1 of Appendix B.

The intersection point x_2 of section 3.4, in which $h_1(x) = h_2(x)$, is located at

$$x_2 = \frac{\sqrt{RT_{x_2}} - t(1-\gamma)(\gamma+10)}{(14t-1)\gamma^2 + (6t+1)\gamma}, \tag{4.3}$$

in which $RT_{x_2} = t(1-\gamma)[(209t-15)\gamma^3 + (141t+10)\gamma^2 + 5(1-10t)\gamma + 100t]$ is the root term of x_2. The parameter domains are $\gamma \in (0,1)$ and $t \geq 5$.

Equivalence class k of section 3.4 is determined to be $l = 19$. The intersection point x_{19}, in which $h_1(x) = h_{19}(x)$ is located at

$$x_{19} = \frac{\sqrt{RT_{x_{19}}} - t(1-\gamma)(4\gamma^2 + 29\gamma + 35)}{2[(8t-2)\gamma^3 + 40t\gamma^2 + (17t+2)\gamma - 5t]}, \tag{4.4}$$

in which $RT_{x_{19}} = t(1-\gamma)[(304t-80)\gamma^5 + (2664t-320)\gamma^4 + 6511t\gamma^3 + (3211t+320)\gamma^2 + (685t+80)\gamma + 1025t]$ is the root term of x_{19}. The denominator of x_{19} is positive if the parameter

4.3 Sequence Length $p = 5$

$l : [...]$	$h_l =$	$c_{11}(l)$	$+2c_{12}(l)x$	$+c_{22}(l)x^2$
$R_1 = 5$				
$1 : [12345]$	$h_1 =$	4	$-\frac{8}{5} \cdot x$	$+\frac{4(4t-1)}{5t} \cdot x^2$
Group A: $R_A = (3\gamma+5)/(\gamma+1)$				
$2 : [12344]$	$h_2 =$	$\frac{6(\gamma+3)}{(3\gamma+5)}$	$-\frac{4\gamma}{(3\gamma+5)} \cdot x$	$+\frac{(-(4t-2)\gamma^2-(4t-2)\gamma+16t-4)}{t(1-\gamma)(3\gamma+5)} \cdot x^2$
Group B: $R_B = (\gamma+5)/(\gamma+1)$				
$k \equiv 19 : [12233]$	$h_{19} =$	$\frac{8(\gamma+2)}{(\gamma+1)(\gamma+5)}$	$+\frac{2(3+\gamma)}{(\gamma+1)(\gamma+5)} \cdot x$	$+\frac{(4\gamma^2+10t\gamma+14t-4)}{t(1-\gamma)(\gamma+1)(\gamma+5)} \cdot x^2$
Group C: $R_C = (4\gamma+5)/(2\gamma+1)$				
$7 : [12333]$	$h_7 =$	$\frac{2(2\gamma+7)}{(4\gamma+5)}$	$+\frac{4(1-\gamma)}{(4\gamma+5)} \cdot x$	$+\frac{(-(2t-2)\gamma^2+2\gamma+14t-4)}{t(1-\gamma)(4\gamma+5)} \cdot x^2$
$12 : [12322]$	$h_{12} =$	$\frac{2(2\gamma+7)}{(4\gamma+5)}$	$-\frac{6}{(4\gamma+5)} \cdot x$	$+\frac{(-(2t-2)\gamma^2+(6t+2)\gamma+14t-4)}{t(1-\gamma)(4\gamma+5)} \cdot x^2$
$20 : [12232]$				
Group D: $R_D = (7\gamma+5)/((\gamma+1)(2\gamma+1))$				
$43 : [11222]$	$h_{43} =$	$\frac{12}{(7\gamma+5)}$	$+\frac{10}{(7\gamma+5)} \cdot x$	$+\frac{((10t-2)\gamma+12t-4)}{t(1-\gamma)(7\gamma+5)} \cdot x^2$
$27 : [12211]$	$h_{27} =$	$\frac{12}{(7\gamma+5)}$	$+0$	$+\frac{((16t-4)\gamma+12t-4)}{t(1-\gamma)(7\gamma+5)} \cdot x^2$
$44 : [11221]$				
Group E: $R_E = (3\gamma+5)/(3\gamma+1)$				
$23 : [12222]$	$h_{23} =$	$\frac{8}{(3\gamma+5)}$	$+\frac{4}{(3\gamma+5)} \cdot x$	$+\frac{((2t+4)\gamma+10t-4)}{t(1-\gamma)(3\gamma+5)} \cdot x^2$
$37 : [12111]$	$h_{37} =$	$\frac{12}{(3\gamma+5)}$	$-\frac{6}{(3\gamma+5)} \cdot x$	$+\frac{((4t-2)\gamma+10t-4)}{t(1-\gamma)(3\gamma+5)} \cdot x^2$
$47 : [11211]$				
$50 : [11121]$				

Table 4.3: Some equivalence classes l, their representative sequences [...] and h_l functions for sequence length $p = 5$.

domain for γ is restricted to the interval $(0.3, 1)$, and $t \geq 5$. Thus, $x_{19} \in (0, 1)$, cf. Proposition 4.

The 36 equivalence classes of $l \in \{2, \ldots, 51\}$ are divided into groups of identical R_u. Equivalence class l belongs to

4 Optimal Designs

group A $\Leftrightarrow l \in L_A = \{2, 3, 5, 6, 10, 14, 38\}$;
group B $\Leftrightarrow l \in L_B = \{8, 9, 11, 13, 15, 16, 19, 21, 25, 32, 39, 42\}$;
group C $\Leftrightarrow l \in L_C = \{7, 12, 17, 22, 31, 35, 48\}$;
group D $\Leftrightarrow l \in L_D = \{24, 26, 27, 34, 36, 43, 49\}$;
group E $\Leftrightarrow l \in L_E = \{23, 37, 51\}$.

As in the case of $p = 3$ or $p = 4$, the coefficients $c_{11}(l)$ are identical within each group, A through E. Thus, the intention is to identify the maximal h_l function of each group by comparing all $c_{12}(l)$ and $c_{22}(l)$ within the group, assuming $x > 0$.

A) $l \in L_A$: The maximal h_l function is $h_2(x)$, because $c_{12}(2) > c_{12}(l)$, and $c_{22}(2) \geq c_{22}(l)$ for all l.

B) The same criteria as in A) hold for h_{19}, which is the maximum of all h_l, $l \in L_B$.

C) In group C, we have $c_{12}(7) > c_{12}(l)$ and $c_{22}(12) > c_{22}(l)$ for all $l \in L_C$. The sequences l with $c_{22}(l) > c_{22}(7)$ have smaller $c_{12}(l)$ and $c_{22}(l)$ than h_{12}. Thus, there are two h_l functions, $l = 7$ and $l = 12$, representing the maximum in this group, dependent on the magnitude of γ.

D) Similar to C), the maximal h_l functions within group D are given by $l = 27$ and $l = 43$, dependent on the magnitude of γ. In this group, $c_{12}(43) > c_{12}(l)$ for all $l \in L_D$ and $c_{22}(43) > c_{22}(l)$ for all $l \in L_D \setminus \{27\}$.

E) Attach $l = 52$ to group E. As a result, the maximal h_l is again represented by two equivalence classes: $l = 23$ and $l = 37$, dependent on the magnitude of γ. This is derived from $c_{11}(23) \geq c_{11}(l)$, $c_{12}(23) > c_{12}(l)$ for all $l \in L_D \cup \{52\}$. Additionally, $c_{22}(23) \geq c_{22}(l)$ for all $l \in \{L_D \cup \{52\}\} \setminus \{37\}$ yielding $h_{37} > h_{23}$ when γ is getting close to 1.

Lemma 7. The abscissa x_{19} is the intersection point of equivalence class functions $h_1(x) = h_{19}(x)$, cf. equation (4.4). For any $t \geq p = 5$, define the parameter $\gamma_{\beta(t)} \in (0.92, 0.93)$ as the γ-value for which $h_2(x_{19}) = h_{19}(x_{19})$.
The maximum of all $h_l(x_{19})$, $l \in \{2, \ldots, 52\}$, is $h_{19}(x_{19})$ for all $\gamma \in (\gamma_{\beta(t)}, 1)$ and all $t \geq 5$.

Proof. Consider the equivalence class functions h_l of Table 4.3 and use the results of A) through E) above. There are just eight equivalence class functions h_l, $l \in \{2, 19, 7, 12, 43, 27, 23, 37\}$, which dominate the h_l functions, $l \in \{2, \ldots, 52\} \setminus \{2, 19, 7, 12, 43, 27, 23, 37\}$, for $x > 0$. Since $x_{19} > 0$ for all $\gamma > 0.3$, we have $h_{19}(x_{19}) = \max\limits_{l \in \{2,\ldots,52\}} h_l(x_{19})$ iff $h_{19}(x_{19}) > h_l(x_{19})$ for all $l \in \{2, 7, 12, 43, 27, 23, 37\}$ and $\gamma_{\beta(t)} < \gamma < 1$. Therefore, analyze

a)

$$(h_{19} - h_2)(x) = -\frac{2(3\gamma^3 + 15\gamma^2 + 25\gamma + 5)}{(\gamma+1)(\gamma+5)(3\gamma+5)} + \frac{2(2\gamma^3 + 15\gamma^2 + 24\gamma + 15)}{(\gamma+1)(\gamma+5)(3\gamma+5)} \cdot x$$
$$-\frac{2[(1-2t)\gamma^4 + (1-14t)\gamma^3 + (-29t-1)\gamma^2 + (-8t-1)\gamma + 5t]}{t(1-\gamma)(\gamma+1)(\gamma+5)(3\gamma+5)} x^2.$$

Substitution of $x = x_{19}$ provides

$$(h_{19} - h_2)(x_{19}) = \frac{a(\gamma) + b(\gamma)\sqrt{RT_{x_{19}}}}{(3\gamma+5)[(8t-2)\gamma^3 + 40t\gamma^2 + (17t+2)\gamma - 5t]^2},$$

with

$a(\gamma) = \;$ $-16(2t^2 + 3t - 1)\gamma^7 + 4(24t^2 - 65t - 4)\gamma^6 + (1286t^2 + 175t - 16)\gamma^5 + 2(109t^2 + 98t + 8)\gamma^4 - 2t(1232t + 51)\gamma^3 + 8t(82t + 3)\gamma^2 + 5t(178t + 3)\gamma - 650t^2$

and

$b(\gamma) = \;$ $8t\gamma^4 + (38t + 3)\gamma^3 - 2(17t + 1)\gamma^2 - (24t + 1)\gamma + 20t.$

The denominator of $(h_{19} - h_2)(x_{19})$ is positive for arbitrary values of γ and t. In order to determine the roots of $a(\gamma) + b(\gamma)\sqrt{RT_{x_{19}}}$, the signs of $a(\gamma)$ and $b(\gamma)$ for $\gamma \in (0.92, 1) \supset (\gamma_{\beta(t)}, 1)$ have to be identified.

a1) The transformation $\gamma = 0.08\gamma' + 0.92$, in which $\gamma' \in (0, 1)$, yields

$a(\gamma') = \;$ $\frac{16(1-\gamma')}{6103515625}[128(2t^2 + 3t - 1)\gamma'^6 + 16(704t^2 + 3581t - 552)\gamma'^5 - 10(154766t^2 - 269901t + 23392)\gamma'^4 - 5(20560424t^2 - 11815289t + 594688)\gamma'^3 - 5(407891082t^2 - 134949577t + 3542184)\gamma'^2 - (15626081286t^2 - 3989533821t + 32802232)\gamma' - (57425437544t^2 - 9922322659t - 53729472)].$

Define $a(\gamma') = \frac{16(1-\gamma')}{6103515625}[a_6\gamma'^6 + \ldots + a_1\gamma' + a_0]$. It follows that $a_i < 0$ for all $0 \leq i \leq 4$ and $t \geq 5$. Furthermore, $a_6\gamma'^6 + a_5\gamma'^5 + a_4\gamma'^4 < (a_6 + a_5 + a_4)\gamma'^4 < 0$ for $\gamma' \in (0, 1)$. Then we conclude that $a(\gamma) < 0$, for $\gamma \in (0.92, 1)$.

a2) Apply the same transformation $\gamma = 0.08\gamma' + 0.92$ with $\gamma' \in (0, 1)$ to $b(\gamma)$ to get

$b(\gamma') = \;$ $\frac{4}{390625}[32t\gamma'^4 + 6(562t + 25)\gamma'^3 + (69692t + 3925)\gamma'^2 + 3(90749t + 7650)\gamma' + 435907t - 27025].$

All coefficients of γ'^i, $0 \leq i \leq 4$, are positive. Hence, $b(\gamma')$ is positive for $\gamma' \in (0, 1)$, such is $b(\gamma)$ for all $\gamma \in (0.92, 1)$ and $t \geq 5$.

4 Optimal Designs

Subtracting $a(\gamma)$ on both sides of $a(\gamma) + b(\gamma)\sqrt{RT_{x_{19}}} > 0$ and squaring both sides of the inequality afterwards is equivalent to

$$b^2(\gamma)RT_{x_{19}} - a^2(\gamma) > 0$$
$$\Leftrightarrow 4(1-\gamma)(8t\gamma^3 - 2\gamma^3 + 40t\gamma^2 + 17t\gamma + 2\gamma - 5t)^2 \cdot g(\gamma) > 0$$
$$\Leftrightarrow g(\gamma) := g_0 + g_1\gamma + \ldots + g_7\gamma^7 > 0,$$

in which

$$\begin{aligned}-g(\gamma) =\ & -16(5t^2+2t+1)\gamma^7 - 8(71t^2-t-2)\gamma^6 - t(698t-31)\gamma^5 + 2t(1241t+ \\ & 44)\gamma^4 - 5t(83t+10)\gamma^3 - 5t(303t+8)\gamma^2 + 5t(121t-1)\gamma + 125t^2.\end{aligned}$$

In order to find the roots of $g(\gamma)$, decompose $-g(\gamma)$ into $G_{03}(\gamma) := -(g_0 + \ldots + g_3\gamma^3)$ and $G_{47}(\gamma) := -(g_4\gamma^4 + \ldots + g_7\gamma^7)$. The mein purpose is to show that $G_{03}(\gamma)$ intercepts once in interval $(0,1)$, and that $G_{47}(\gamma)$ is positive and increasing in γ.

a3) The second derivative of $G_{03}(\gamma)$ is $G_{03}''(\gamma) = -(2g_2 + 6g_3\gamma) < 0$ for all γ and all t because g_2 and g_3 are positive. $G_{03}''(\gamma) < 0$ implies that $G_{03}(\gamma)$ is concave, i.e., the local minima of $G_{03}(\gamma)$ are represented by the end points of $(0,1)$. Hence, $G_{03}(\gamma)$ cannot have more than two roots in the interval $(0,1)$. But as the function values of $G_{03}(\gamma)$ for γ getting close to 0 or 1 have different signs, there is just one root possible for $G_{03}(\gamma)$.

a4) For all $t \geq 5$, we have $-g_4\gamma^4 > 0$ and all $-g_i\gamma^i$, $i = 5,6,7$, are negative. Furthermore, $0 < -(g_4+g_i)\gamma^i < -g_4\gamma^4 - g_i\gamma^i$ and $0 < -(g_4 + \ldots + g_i)$ for all $i = 5,6,7$. Thus, $G_{47}(\gamma) > 0$ for all t and all γ.

The first derivative of $G_{47}(\gamma)$ is $G_{47}'(\gamma) = -\sum_{i=4}^{7} ig_i\gamma^{i-1}$. The coefficient $-g_4 > 0$ and the other expressions $-g_i$, $i = 5,6,7$, are negative. In general, $\gamma^{i-1} > \gamma^i$, $\forall i \in \mathbb{N}$, and finally $\sum_{i=4}^{j} ig_i > 0$ for all $j = 5,6,7$. The conclusion is that $G_{47}'(\gamma)$ is positive for all $t \geq 5$ and all $\gamma \in (0,1)$. As $G_{47}'(\gamma)$ is the slope of $G_{47}(\gamma)$, $G_{47}(\gamma)$ is monotonous and increasing in γ.

Combine the results of a3) and a4) to conclude that there are only two roots possible to exist for $-g(\gamma)$ in the interval $(0,1)$, because $G_{03}(\gamma)$ is monotonous decreasing iff $G_{03}(\gamma) < 0$, and $G_{47}(\gamma) > 0$ is increasing and positive. Iff $G_{03}(\gamma) > 0$, the function $-g(\gamma)$ is positive as $G_{47}(\gamma) > 0$ for all $\gamma \in (0,1)$.

The function $-g(\gamma)$ is continuously extendible on $[0,1]$. Since $-g(0) = 125t^2$ is positive and $-g(1) = -64t^2$ is negative, $-g(\gamma)$ can only have one root $\gamma_{\beta(t)}$ in the interval $(0,1)$.

Using $-g(0.92) > 0$ and $-g(0.93) < 0$, it follows that $\gamma_{\beta(t)} \in (0.92, 0.93)$ and all $t \geq 5$. To summarize, $(h_{19} - h_2)(x_{19}) > 0$ for all $\gamma \in (\gamma_{\beta(t)}, 1)$. Thus, $h_{19}(x_{19}) > h_l(x_{19}), l \in L_A$.

b) through g) is treated in Appendix B, section B.1.2.

By a) through g), it follows that $h_{19}(x_{19}) = \max_l h_l(x_{19})$ for all $2 \leq l \leq 52$ iff $\gamma > \gamma_\beta$. □

Lemma 8. The abscissa x_2 is the intersection point of equivalence class functions $h_1(x) = h_2(x)$, cf. equation (4.3). Similar to Lemma 7, consider the same parameter $\gamma_{\beta(t)} \in (0.92, 0.93)$ as the γ-value for which $h_2(x_{19}) = h_{19}(x_{19})$, and $h_2(x_2) = h_{19}(x_2)$ as well, i.e., $x_2 = x_{19}$, for any $t \geq p = 5$.
The maximum of all $h_l(x_2)$, $l \in \{2, \ldots, 52\}$, is $h_2(x_2)$ for all $\gamma \in (0, \gamma_{\beta(t)})$ and all $t \geq 5$.

Proof. Recall the equivalence class functions h_l of Table 4.3 and use the results of A) through E) above. There are just eight equivalence class functions, h_l, $l \in \{2, 7, 12, 19, 43, 27, 23, 37\}$, which can exceed the h_l functions, $l \in \{2, \ldots, 52\} \setminus \{2, 7, 12, 19, 43, 27, 23, 37\}$, for $x > 0$. Since $x_2 > 0$, $h_2(x_2) = \max_{l \in \{2,\ldots,52\}} h_l(x_2)$ iff $h_2(x_2) > h_l(x_2)$ for all $l \in \{7, 12, 19, 43, 27, 23, 37\}$ and $0 < \gamma < \gamma_{\beta(t)}$. For this purpose, analyze

a)

$$(h_2 - h_{19})(x) = \frac{2(3\gamma^3 + 15\gamma^2 + 25\gamma + 5)}{(\gamma + 1)(\gamma + 5)(3\gamma + 5)} - \frac{2(2\gamma^3 + 15\gamma^2 + 24\gamma + 15)}{(\gamma + 1)(\gamma + 5)(3\gamma + 5)}x$$
$$+ \frac{2[(1 - 2t)\gamma^4 + (1 - 14t)\gamma^3 + (-29t - 1)\gamma^2 + (-8t - 1)\gamma + 5t]}{t(1 - \gamma)(\gamma + 1)(\gamma + 5)(3\gamma + 5)}x^2.$$

Inserting $x = x_2$ provides

$$(h_2 - h_{19})(x_2) = \frac{a(\gamma) + b(\gamma)\sqrt{RT_{x_2}}}{\gamma^2(\gamma + 1)(\gamma + 5)(14t\gamma - \gamma + 6t + 1)^2},$$

in which
$$a(\gamma) = 8(12t^2 + 13t - 1)\gamma^6 - 2(142t^2 + 27t - 8)\gamma^5 - 4(231t^2 + 19t + 2)\gamma^4 + 16t(95t + 3)\gamma^3 - 4t(7t + 3)\gamma^2 - 10t(78t + 1)\gamma + 400t^2$$

and
$$b(\gamma) = -16t\gamma^4 - 2(38t + 3)\gamma^3 + 4(17t + 1)\gamma^2 + 2(24t + 1)\gamma - 40t.$$

The denominator of $(h_2 - h_{19})(x_2)$ is positive for all $\gamma \in (0, 1)$ and $t \geq 5$. Next we have to examine the roots of $a(\gamma) + b(\gamma)\sqrt{RT_{x_2}}$. The signs of $a(\gamma)$ and $b(\gamma)$ need to be determined for all $\gamma \in (0, 0.93) \supset (0, \gamma_{\beta(t)})$.

4 Optimal Designs

a1) Put $a(\gamma) = a_6\gamma^6 + \ldots + a_1\gamma + a_0$ and define $A_{36} = \sum_{i=3}^{6} a_i\gamma^i$ and $A_{02} = a_0 + a_1\gamma + a_2\gamma^2$.
We have $a_6 > 0$ and $a_5\gamma^5 + a_4\gamma^4 + a_3\gamma^3 > (a_5 + a_4 + a_3)\gamma^5 > 0$. Hence, $A_{36} > 0$ for all $\gamma \in (0, 0.93)$ and $t \geq 5$.
One root of A_{02} lies in the interval $(0, 0.93)$ and is located at $\gamma_{root} = \frac{5\sqrt{6532t^2+348t+1}-5(78t+1)}{4(7t+3)}$ with $A_{02} > 0$ for all $\gamma < \gamma_{root}$. Some simple calculus confirms that $\gamma_{root} > 0.5$. Thus $a(\gamma)$ is positive for all $\gamma \in (0, 0.5]$.
Next, transform $\gamma = 0.5 + 0.43\gamma'$ such that $\gamma \in (0.5, 0.93)$ and $\gamma' \in (0, 1)$. The value of

$$\begin{aligned}a(\gamma') = & \ [6321363049(12t^2 + 13t - 1)\gamma'^6 + 3675211075(2t^2 + 129t - 4)\gamma'^5 - \\ & \ 4273501250(1274t^2 - 179t + 2)\gamma'^4 - 9938375000(798t^2 - 21t - \\ & \ 4)\gamma'^3 + 11556250000(1202t^2 - 48t + 1)\gamma'^2 - 6718750000(1606t^2 + \\ & \ 171t + 4)\gamma' + 7812500000(2046t^2 - 109t - 2)]/125000000000\end{aligned}$$

can be written as $a(\gamma') = (a'_6\gamma'^6 + \ldots + a'_1\gamma' + a'_0)/125000000000$. Furthermore, $a'_1\gamma' + a'_0 > (a'_1 + a'_0)\gamma' > 0$, $a'_4\gamma'^4 + a'_3\gamma'^3 + a'_2\gamma'^2 > (a'_4 + a'_3 + a'_2)\gamma'^4 > 0$, $a'_5 > 0$ and $a'_6 > 0$. It follows that $a(\gamma') > 0$ for all $\gamma' \in (0, 1)$ and, therefore, $a(\gamma) > 0$ for all $\gamma \in (0.5, 0.93)$ and $t \geq 5$.
Hence, $a(\gamma)$ is positive for all $\gamma \in (0, 0.93)$.

a2) The second derivative of $b(\gamma)$ is $b''(\gamma) = -192t\gamma^2 - 12(38t + 3)\gamma + 8(17t + 1)$. Its root in range $(0, 1)$ is located at $y_{root} = \frac{1}{96t}[\sqrt{3(6508t^2 + 812t + 27)} - 3(38t + 3)]$ such that $b''(\gamma)$ is positive for all $\gamma \in (0, \gamma_{root})$ and negative otherwise. Some calculus shows $y_{root} < 0.3$. The sign of $b''(\gamma)$ implies that $b(\gamma)$ is concave on the interval $(0, \gamma_{root})$ and convex on the interval $(\gamma_{root}, 0.93)$. Since $b(\gamma)$ is negative for $\gamma = \gamma_{root}$ and $\gamma = 0.93$, it follows that $b(\gamma)$ is negative for all $\gamma \in [0.3, 0.93)$.
Analyze $b(\gamma)$ in the interval $(0, 0.3)$ by transforming $\gamma = 0.3\gamma'$, such that $\gamma \in (0, 0.3)$ and $\gamma' \in (0, 1)$. Then

$$\begin{aligned}b(\gamma') = & \ [(-324t\gamma'^4 - 135(38t + 3)\gamma'^3 + 900(17t + 1)\gamma'^2 + 1500(24t + 1)\gamma' - \\ & \ 100000t]/2500 := (b_4\gamma'^4 + \ldots + b_1\gamma' + b_0)/2500.\end{aligned}$$

Observe that b_4, b_3, and $b_2\gamma'^2 + b_1\gamma' + b_0 < (b_2 + b_1 + b_0)$ are negative. It follows that $b(\gamma')$ is negative for all $\gamma' \in (0, 1)$ and $\gamma \in (0, 0.3)$.
Finally, $b(\gamma)$ is negative for all $\gamma \in (0, 0.93)$.

As a consequence of a1) and a2), $a(\gamma) + b(\gamma)\sqrt{RT_{x_2}} > 0$ is equivalent to

$$a^2(\gamma) - b^2(\gamma)RT_{x_2} > 0$$
$$\Leftrightarrow 4\gamma^2(1-\gamma)[(14t-1)\gamma + 6t + 1]^2 \cdot g^*(\gamma) > 0$$
$$\Leftrightarrow g^*(\gamma) := g_0^* + g_1^*\gamma + \ldots + g_7^*\gamma^7 > 0,$$

in which $g^*(\gamma) \equiv -g(\gamma)$ in the proof of Lemma 7 a). It is required that $g^*(\gamma)$ and $-g(\gamma)$ have the same root at $\gamma_{\beta(t)}$, the specific parameter, in which $x_2 = x_{19}$.

Hence, $(h_2 - h_{19})(x_2) > 0$ for all $\gamma \in (0, \gamma_{\beta(t)})$, and $h_2(x_2) > h_l(x_2)$, for all $l \in L_B$.

b) Knowing that $h_{19}(x) > h_l(x)$ for all $l \in \{7, 12, 23, 37, 43\}$, $\gamma \in (0, 1)$ and $x \in (0, 1)$, cf. proof of Lemma 7 properties b) through d), f) and g), and using $h_2(x_2) > h_{19}(x_2)$ for all $\gamma \in (0, \gamma_{\beta(t)})$, it follows that $h_2(x_2) > h_l(x_2)$ for all $l \in \{7, 12, 23, 37, 43\}$.

c) The analysis of $(h_2 - h_{27})(x)$ continues in Appendix B, section B.1.2.

Finally, $h_2(x_2) = \max_l h_l(x_2)$ for all $l \in \{2, 7, 12, 19, 43, 27, 23, 37\}$ iff $\gamma \in (0, \gamma_\beta)$. Lemma 8 follows. □

Lemma 9. For any $t \geq p = 5$, the parameters $0 < \gamma_{\alpha 1} < \gamma_{\alpha 2} < 1$ are given by

$$\gamma_{\alpha 1/\alpha 2} = \frac{\mp\sqrt{51076t^4 - 64428t^3 + 28761t^2 - 5560t + 400} + 226t^2 - 99t + 10}{2(246t^2 - 119t + 15)}.$$

Let $\gamma \in (\gamma_1, \gamma_2)$, and observe that $x_{min} = t/(4t-1)$ is the abscissa of the minimum of equivalence class function $h_1(x)$, then, $h_1(x_{min}) = \max_{l \in \{1,\ldots,52\}} h_l(x_{min})$.

Proof. As $x_{min} > 0$, we can use the results of A) through E) above, i.e., one of the h_l, $l \in \{2, 7, 12, 19, 43, 27, 23, 37\}$ is maximal within all other h_l functions, $l \in \{2,\ldots,52\} \setminus \{2, 7, 12, 19, 43, 27, 23, 37\}$, for any $x > 0$. Using the results of the proof of Lemma 7, i.e., $h_{19} = \max_{l \in \{7,12,23,37,43\}}(h_l)$ for all $x \in (0,1)$, it remains to verify whether $h_1(x_{min}) > h_l(x_{min})$ for $l \in \{2, 19, 27\}$, and it follows that $h_1(x_{min}) = \max_{l=1,\ldots,52} h_l(x_{min})$. Therefore, consider the first difference of interest:

a)
$$(h_2 - h_1)(x_{min}) = \frac{2(246t^2 - 119t + 15)\gamma^2 - 2(226t^2 - 99t + 10)\gamma + 10(4t - 1)}{5(4t-1)^2(1-\gamma)(3\gamma + 5)}.$$

The solution of the quadratic equation $2(246t^2 - 119t + 15)\gamma^2 - 2(226t^2 - 99t + 10)\gamma + 10(4t-1) = 0$ for all $\gamma \in (0, 1)$ and all $t \geq 5$ yields

$$(h_2 - h_1)(x_{min}) \begin{cases} < 0 & \Leftrightarrow \gamma \in (\gamma_{\alpha 1}, \gamma_{\alpha 2}) \\ \geq 0 & \Leftrightarrow \gamma \notin (\gamma_{\alpha 1}, \gamma_{\alpha 2}) \end{cases}.$$

Hence, $h_1(x_{min}) > h_2(x_{min})$ for $\gamma \in (\gamma_{\alpha 1}, \gamma_{\alpha 2})$.

b) Consider now
$$(h_{19} - h_1)(x_{min}) = \frac{2 \cdot g(\gamma)}{5(4t-1)^2(1-\gamma^2)(\gamma+5)},$$
in which $g(\gamma) = 2(4t-1)(19t-5)\gamma^3 + 5(84t^2-43t+6)\gamma^2 - (487t^2-248t+30)\gamma - 5(5t^2-9t+2)$.

The denominator of $(h_{19} - h_1)(x_{min})$ is positive for all $\gamma \in (0,1)$ and t. Let us consider the numerator $2 \cdot g(\gamma)$ of $(h_{19} - h_1)(x_{min})$. Differentiate $g(\gamma)$ twice to obtain $g''(\gamma) = 12(4t-1)(19t-5)\gamma + 10(84t^2-43t+6) > 0$ for all $t \geq 5$ and $\gamma \in (0,1)$. A positive curvature means that $g(\gamma)$ is convex and may have up to two roots in the interval $(0,1)$. By the continuity of $g(\gamma)$, it follows that the function has exactly one root at $\gamma_{root} \in (0,1)$ because $g(\gamma \searrow 0)$ is negative and $g(\gamma \nearrow 1)$ is positive. Thus,

$$(h_{19} - h_1)(x_{min}) \begin{cases} < 0 & \Leftrightarrow \gamma < \gamma_{root} \\ \geq 0 & \Leftrightarrow \gamma \geq \gamma_{root} \end{cases}.$$

Further, the value of $g(\gamma)$ at $\gamma_{\alpha 2}$ is

$$g(\gamma_{\alpha 2}) = \frac{5(Z\sqrt{W} - V)}{2(246t^2 - 119t + 15)^3},$$

in which
$Z = 328516t^6 - 297984t^5 + 75149t^4 + 10659t^3 - 8790t^2 + 1600t - 100,$
$W = 51076t^4 - 64428t^3 + 28761t^2 - 5560t + 400$
and
$V = 74624744t^8 - 113201004t^7 + 64841794t^6 - 13812609t^5 - 2042961t^4 + 1810990t^3 - 421800t^2 + 45900t - 2000.$

Remodel $g(\gamma_{\alpha 2})$ properly to get $g(\gamma_{\alpha 2}) < 0$. However, this inequality implies that $\gamma_{\alpha 2} < \gamma_{root}$.

Hence, $h_1(x_{min}) > h_{19}(x_{min})$ for all $\gamma \in (\gamma_{\alpha 1}, \gamma_{\alpha 2})$.

c) The last difference to examine is

$$(h_{27} - h_1)(x_{min})$$
$$= \frac{4[7(4t-1)(19t-5)\gamma^2 - (4t-1)(93t-25)\gamma - 5(25t^2 - 14t + 2)]}{5(4t-1)^2(1-\gamma)(7\gamma+5)}.$$

Again, solving the quadratic equation $7(4t-1)(19t-5)\gamma^2 - (4t-1)(93t-25)\gamma - 5(25t^2 - 14t+2) = 0$ yields

$$(h_{27} - h_1)(x_{min}) \begin{cases} <0 & \Leftrightarrow \gamma \in (\gamma_3, \gamma_4) \\ \geq 0 & \Leftrightarrow \gamma \notin (\gamma_3, \gamma_4) \end{cases}$$

with

$$\gamma_{3/4} = \frac{\mp\sqrt{(4t-1)(101096t^3 - 81989t^2 + 22270t - 2025)} + (4t-1)(93t-25)}{14(4t-1)(19t-5)}.$$

Some simple calculus shows that $\gamma_3 < 0$ and $\gamma_4 > 0.93$. Hence, for all $\gamma \in (\gamma_{\alpha 1}, \gamma_{\alpha 2}) \subset (0, 0.93)$, the difference $(h_{27} - h_1)(x_{min})$ is negative which is equivalent to $h_1(x_{min}) > h_{27}(x_{min})$ for all $\gamma \in (\gamma_{\alpha 1}, \gamma_{\alpha 2})$.

Combining a), b) and c), $h_1(x_{min}) = \max_l h_l(x_{min})$ for all $\gamma \in (\gamma_{\alpha 1}, \gamma_{\alpha 2})$ and all $l \in \{2, 19, 27\}$, and thus, for all $1 \leq l \leq 52$. Lemma 9 follows. $\square$

Theorem 3. For any $t \geq p = 5$ and $\gamma \in (0, 1)$, consider the parameters $0 < \gamma_{\alpha_1} < \gamma_{\alpha_2} < \gamma_{\beta(t)} < 1$ as in Lemmas 8 and 9. Furthermore, the proportion $\alpha(\gamma) \in [0, 1]$ is given by

$$\alpha(\gamma) = \frac{2(1-\gamma)(3\gamma+5)[10t(t\gamma^2 + (1-3t)\gamma + 4t-1) - (4t-1)\sqrt{RT_{x_2}}]}{\gamma[(14t-1)\gamma + 6t+1]\sqrt{RT_{x_2}}}$$

iff $\gamma \in (0, \gamma_{\alpha 1}] \overset{\bullet}{\cup} [\gamma_{\alpha 2}, \gamma_{\beta(t)}]$ and $\alpha(\gamma) := 0$ iff $\gamma \in (\gamma_{\alpha 1}, \gamma_{\alpha 2})$ or $\gamma \in (\gamma_{\beta(t)}, 1)$. Another proportion $\beta(\gamma) \in [0, 1]$ is given by

$$\beta(\gamma) = \frac{2(1-\gamma^2)(\gamma+5)[5t((5-4t)\gamma^2 + (2t+2)\gamma + 26t - 7) - (4t-1)\sqrt{RT_{x_{19}}}]}{[(8t-2)\gamma^3 + 40t\gamma^2 + (17t+2)\gamma - 5t]\sqrt{RT_{x_{19}}}}$$

iff $\gamma \in [\gamma_{\beta(t)}, 1)$ and $\beta(\gamma) := 0$ otherwise.
The optimality results are as follows:

a) For all $\gamma \in (0, \gamma_{\beta(t)})$, an approximate design d^* is optimal iff $(1 - \alpha(\gamma)) \cdot 100\%$ of its sequences are selected from class 1 with representative sequence $[1, 2, 3, 4, 5]$ and $\alpha(\gamma) \cdot 100\%$ of its sequences from class 2 with representative sequence $[1, 2, 3, 4, 4]$.

b) For all $\gamma \in (\gamma_{\beta(t)}, 1)$, an approximate design d^* is optimal iff $(1 - \beta(\gamma)) \cdot 100\%$ of its sequences are selected from class 1 and $\beta(\gamma) \cdot 100\%$ of its sequences from class 19 with representative sequence $[1, 2, 2, 3, 3]$.

c) Iff $\gamma = \gamma_{\beta(t)}$, define proportion $\varphi \in [0, 1]$. An approximate design d^* is optimal iff $\varphi \cdot 100\%$ of the sequences are arranged as in a) and $(1 - \varphi) \cdot 100\%$ of the sequences are arranged as in b).

4 Optimal Designs

Proof. Theorem 3 claims that x_{d^*} of Proposition 2, in which the $\min_{x} \max_{l} h_l(x)$ is realized, is either x_2, x_{19}, $x_2 = x_{19}$, or x_{min}, the x-coordinate of the minimum of h_1. In order to verify this conjecture, eight properties and formulas need to be derived:

1. For all $\gamma \notin (\gamma_{\alpha 1}, \gamma_{\alpha 2})$ and $\gamma < \gamma_{\beta(t)}$: $h_2(x_2) > h_l(x_2)$ for all $3 \leq l \leq 52$.

2. For all $\gamma \notin (\gamma_{\alpha 1}, \gamma_{\alpha 2})$: $\text{sign}\, h'_1(x_2) \neq \text{sign}\, h'_2(x_2)$.

3. For all $\gamma \in (\gamma_{\alpha 1}, \gamma_{\alpha 2})$: $h_1(x_{min}) > h_l(x_{min})$ for all $2 \leq l \leq 52$.

4. The formula of $\alpha(\gamma)$.

5. For all $\gamma \in (\gamma_{\beta(t)}, 1)$: $h_{19}(x_{19}) > h_l(x_{19})$ for all $2 \leq l \leq 18$ and $20 \leq l \leq 52$.

6. For all $\gamma \in (\gamma_{\beta(t)}, 1)$: $\text{sign}\, h'_1(x_{19}) \neq \text{sign}\, h'_{19}(x_{19})$.

7. The formula of $\beta(\gamma)$.

8. For $\gamma = \gamma_{\beta(t)}$: $x^* := x_2 = x_{19}$ and $h_2(x^*) = h_{19}(x^*) > h_l(x^*)$ for all $3 \leq l \leq 18$ and $20 \leq l \leq 52$.

Statements 1, 3 and 5 follow from Lemmas 8, 9 and 7, respectively. Statement 8 is a consequence of Lemmas 7 and 8. ✓

As required in point 4, the proportion $\alpha(\gamma) \in (0, 1)$ of equivalence class 2 sequences has to be determined. For this purpose, use equation (2.8) and denote $\alpha(\gamma, x)$ as α, then $q_{d^*}(x) = \alpha n h_2(x) + (1 - \alpha) n h_1(x)$. The proportion $\alpha(\gamma, x)$ is derived by setting

$$0 \stackrel{!}{=} \frac{\partial q_{d^*}}{\partial x} = \alpha \left(\frac{8}{5} - \frac{4\gamma}{3\gamma + 5} + \frac{4(1-2t)\gamma^2 + 4(1-2t)\gamma + 8(4t-1)}{t(1-\gamma)(3\gamma + 5)} x - \frac{8(4t-1)}{5t} x \right)$$
$$- \frac{8}{5} + \frac{8(4t-1)}{5t} x$$
$$\Leftrightarrow \alpha(\gamma, x) = \frac{2[(1-4t)x + t](1-\gamma)(3\gamma + 5)}{x\gamma[(14t-1)\gamma + 6t + 1] + t(1-\gamma)(\gamma + 10)}.$$

Substitution of $x = x_2$ into the formula of $\alpha(\gamma, x)$ provides $\alpha(\gamma)$. A graphical presentation of $\alpha(\gamma)$ for different t is displayed in Figure 4.3. ✓

Again, as in the cases $p = 3$ and $p = 4$, it is necessary to consider the condition $\text{sign}\, h'_1(x_2) \neq \text{sign}\, h'_2(x_2)$. Proportion $\alpha(\gamma)$ is negative or exceeds 1 iff $\text{sign}\, h'_1(x_2) = \text{sign}\, h'_2(x_2)$. Thus, for the purpose of proving property 2, it is sufficient to analyze if $\alpha(\gamma)$ or $\alpha(\gamma, x)$, respectively, is nonnegative and less or equal to 1 in the described domains of the parameters t and γ. In this manner, observe that

$$\alpha(\gamma, x = x_2) \stackrel{!}{=} 0 \Leftrightarrow (1 - 4t)x_2 + t = 0$$
$$\stackrel{(4.3)}{\Leftrightarrow} t\gamma[(14t - 1)\gamma + 6t + 1] \cdot$$
$$\cdot [(246t^2 - 119t + 15)\gamma^2 - (226t^2 - 99t + 10)\gamma + 5(4t - 1)] = 0$$
$$\Leftrightarrow \gamma \in \{\gamma_{\alpha 1}, \gamma_{\alpha 2}, -(6t + 1)/(14t - 1), 0\}.$$

The last two elements $-(6t + 1)/(14t - 1)$ and 0 are not in the required domain of γ. Since $t\gamma[(14t-1)\gamma+6t+1] > 0$ for all $\gamma > 0$, $t > 0$, and $[(246t^2 - 119t + 15)\gamma^2 - (226t^2 - 99t + 10)\gamma + 5(4t - 1)]$ is a convex parabola in γ, proportion $\alpha(\gamma) = \alpha(\gamma, x = x_2) \geq 0$ iff $\gamma \notin (\gamma_{\alpha 1}, \gamma_{\alpha 2})$. We get $\alpha(\gamma) \leq 1$ iff its numerator $A - B\sqrt{RT_{x_2}}$ is less or equal to its denominator $C\sqrt{RT_{x_2}}$, whereas
$$A = 2(1 - \gamma)(3\gamma + 5) \cdot 10t(t\gamma^2 + (1 - 3t)\gamma + 4t - 1),$$
$$B = 2(1 - \gamma)(3\gamma + 5)(4t - 1) \text{ and}$$
$$C = \gamma[(14t - 1)\gamma + 6t + 1].$$
All values A, B, C are positive since $\gamma \in (0, 1)$ and $t \geq 5$.

$$\alpha(\gamma) \leq 1 \Leftrightarrow A - B\sqrt{RT_{x_2}} \leq C\sqrt{RT_{x_2}}$$
$$\Leftrightarrow \ldots \Leftrightarrow A^2 - RT_{x_2}(B + C)^2 \leq 0$$
$$\Leftrightarrow 25t(1 - \gamma)\gamma[(14t - 1)\gamma + 6t + 1] \cdot f(\gamma) \leq 0$$
$$\Leftrightarrow f(\gamma) = f_5\gamma^5 + \ldots f_1\gamma + f_0 \leq 0$$

The function $f(\gamma)$ is defined by
$$f(\gamma) = -(70t^2 - 59t + 15)\gamma^5 - (92t^2 - 118t + 35)\gamma^4 + (366t^2 - 253t + 35)\gamma^3 +$$
$$(436t^2 - 404t + 75)\gamma^2 - (640t^2 - 320t + 40)\gamma - (320t^2 - 160t + 20).$$
As $\gamma^i > \gamma^{i+1}$, $i \in \{0, 1, 2\}$, and f_5, f_4 are negative, we obtain that $f(\gamma) < f_5\gamma^5 + f_4\gamma^4 + (f_3 + f_2 + f_1 + f_0)\gamma = f_5\gamma^5 + f_4\gamma^4 - (158t12 + 177t - 50)\gamma < 0$. Hence, $\alpha(\gamma) \leq 1$ for all $\gamma \in (0, 1)$ and $t \geq 5$.
Since $0 \leq \alpha(\gamma) \leq 1$, it follows that $\text{sign } h'_1(x_2) \neq \text{sign } h'_2(x_2)$ for all $\gamma \notin (\gamma_{\alpha 1}, \gamma_{\alpha 2})$. ✓

The proportion $\beta(\gamma) \in (0, 1)$ of equivalence class 19 sequences has to be determined in property 7. For this purpose, use equation (2.8) and put $\beta(\gamma, x) = \beta$. Then, $q_{d^*}(x) = \beta n h_{19}(x) +$

4 Optimal Designs

$(1-\beta)nh_1(x)$. The proportion $\beta(\gamma, x)$ is derived by setting

$$0 \stackrel{!}{=} \frac{\partial q_{d^*}}{\partial x} = \beta \left(\frac{8}{5} + \frac{2(\gamma+3)}{(\gamma+1)(\gamma+5)} + \frac{4[2\gamma^2 + 5t\gamma + 7t - 2]}{t(1-\gamma^2)(\gamma+5)} x - \frac{8(4t-1)}{5t} x \right)$$
$$- \frac{8}{5} + \frac{8(4t-1)}{5t} x$$

$$\Leftrightarrow \beta(\gamma, x) = \frac{4(t + x - 4tx)(1-\gamma^2)(\gamma+5)}{2[(8t-2)\gamma^3 + 40t\gamma^2 + (17t+2)\gamma - 5t]x + t(1-\gamma)(4\gamma^2 + 29\gamma + 35)}.$$

Substitution of $x = x_{19}$ in the formula of $\beta(\gamma, x)$ provides $\beta(\gamma)$. A graphical presentation of $\beta(\gamma)$ for different t is displayed in Figure 4.3. ✓

The condition to achieve that $\beta \in (0, 1)$ is given by sign $h_1'(x_{19}) \neq$ sign $h_{19}'(x_{19})$. Proportion $\beta(\gamma)$ is negative or exceeds 1 iff sign $h_1'(x_{19}) =$ sign $h_{19}'(x_{19})$. However, a value of β which is not in the interval $[0, 1]$ is not valid for an equivalence class proportion. Thus, to prove property 6, it is sufficient to analyze if $\beta(\gamma)$, or $\beta(\gamma, x)$ respectively, is nonnegative and less or equal to 1 in the corresponding domains of the parameters t and γ. For this purpose, observe that

$$\beta(\gamma, x = x_{19}) \stackrel{!}{\geq} 0 \Leftrightarrow (1 - 4t)x_{19} + t \geq 0$$
$$\stackrel{(4.4)}{\Leftrightarrow} \frac{4t \cdot a(\gamma) \cdot b(\gamma)}{(4t-1)^2} > 0,$$

in which
$a(\gamma) = 2(4t-1)\gamma^3 + 40t\gamma^2 + (17t+2)\gamma - 5t$
and
$b(\gamma) = 2(4t-1)(19t-5)\gamma^3 + 5(84t^2 - 43t + 6)\gamma^2 - (487t^2 - 248t + 30)\gamma - 5(5t^2 - 9t + 2).$

Proportion $\beta(\gamma)$ is positive iff $a(\gamma)$ and $b(\gamma)$ are both positive or both negative for all $\gamma \in (\gamma_{\beta(t)}, 1)$.

The function $a(\gamma)$ is equal to a factor of the denominator of x_{19}, which is positive for all $\gamma \in (0.3, 1)$, cf. Proposition 4. Since $(\gamma_{\beta(t)}, 1)$ is a subset interval of $(0.92, 1)$, apply the transformation $\gamma = 0.08\gamma' + 0.92$ to $b(\gamma)$, in which $\gamma' \in (0, 1)$. This transformation leads to

$b(\gamma') = [16(4t-1)(19t-5)\gamma'^3 + 4(20988t^2 - 10757t + 1440)\gamma'^2 + 34(24697t^2 - 12708t + 1860)\gamma' + 2(6317t^2 + 237862t - 34540)]/15625.$

Observe that all coefficients of γ'^i of $b(\gamma')$ are positive, $0 \leq i \leq 3$. Thus, $b(\gamma')$ is positive for all $\gamma' \in (0, 1)$. Hence, $\beta(\gamma)$ is positive for all $\gamma \in (\gamma_{\beta(t)}, 1)$.
Next to verify is whether proportion $\beta(\gamma) \leq 1$ holds. This condition is fulfilled iff the numerator

of $\beta(\gamma)$ does not exceed the denominator of $\beta(\gamma)$, i.e.,

$$\beta(\gamma) \leq 1 \Leftrightarrow \beta(\gamma, x) \leq 1$$
$$\Leftrightarrow 4(t + x - 4tx)(1 - \gamma^2)(\gamma + 5)$$
$$\leq 2[(8t - 2)\gamma^3 + 40t\gamma^2 + (17t + 2)\gamma - 5t]x + t(1 - \gamma)(4\gamma^2 + 29\gamma + 35)$$
$$\Leftrightarrow -2[2\gamma^2 + 5t\gamma + (7t - 1)]x - t(3 - 2\gamma - \gamma^2) \leq 0.$$

Since $\gamma \in (0, 1)$, it follows that $-t(3 - 2\gamma - \gamma^2) \leq 0$. Proportion $\beta(\gamma)$ is defined for all $\gamma \in [\gamma_{\beta(t)}, 1)$, whereas $0.3 \ll \gamma_{\beta(t)}$. Hence, $x = x_{19} > 0$ and $-2[2\gamma^2 + 5t\gamma + (7t - 1)]x < 0$. Consequently, $\beta(\gamma) = \beta(\gamma, x = x_{19}) \leq 1$.
Thus, sign $h'_1(x_{19}) \neq$ sign $h'_{19}(x_{19})$. ✓

Theorem 3 follows with statements 1. - 8. □

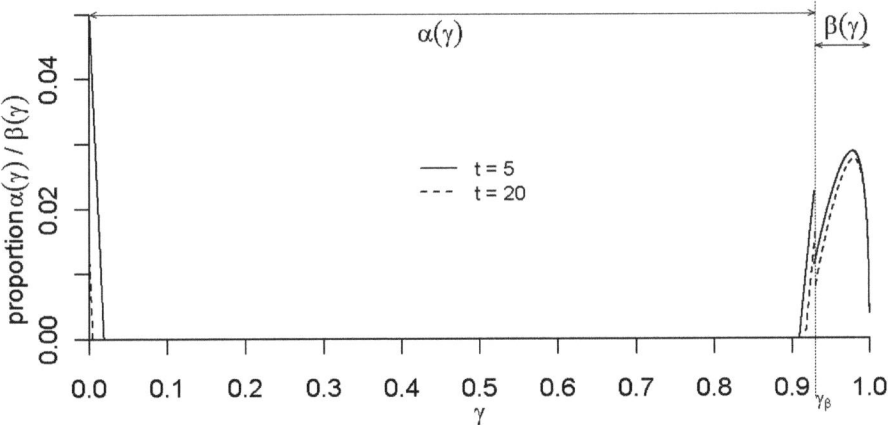

Figure 4.3: Sequence proportions $\alpha(\gamma)$ and $\beta(\gamma)$ of equivalence classes 2 and 19, respectively, for an approximate optimal design with $p = 5$ periods.

4.4 Sequence Length $p = 6$

In the case of 6 periods for each sequence, there are 112 different equivalence classes for $l = 1, \ldots, 203$. Sixteen relevant equivalence classes are listed in table 4.4. For completeness,

4 Optimal Designs

the set of all equivalence classes is given in Table B.2 of Appendix B.

$l : [...]$	$h_l = c_{11}(l)$	$+2c_{12}(l)x$	$+c_{22}(l)x^2$
$R_1 = 6$			
1 : [123456]	5	$-\frac{5}{3}x$	$+\frac{5(5t-1)}{6t}x^2$
Group A: $R_A = 2(2\gamma+3)/(\gamma+1)$			
2 : [123455]	$\frac{2(3\gamma+7)}{(2\gamma+3)}$	$-\frac{3\gamma}{(2\gamma+3)}x$	$+\frac{((3-9t)\gamma^2+(2-6t)\gamma+25t-5)}{2t(1-\gamma)(2\gamma+3)}x^2$
Group B: $R_B = 2(\gamma+3)/(\gamma+1)$			
28 : [123344]	$\frac{(\gamma^2+10\gamma+13)}{((\gamma+1)(\gamma+3))}$	$+\frac{(4-\gamma^2+\gamma)}{(\gamma+1)(\gamma+3)}x$	$+\frac{(-(t-1)\gamma^3+(5-5t)\gamma^2+(15t-1)\gamma+23t-5)}{(2t(1-\gamma)(\gamma+1)(\gamma+3))}x^2$
Group C: $R_C = 6/(\gamma+1)$			
170 : [112233]	$\frac{4}{(\gamma+1)}$	$+\frac{8}{3(\gamma+1)}x$	$+\frac{((9t-1)\gamma+21t-5)}{6t(1-\gamma)(\gamma+1)}x^2$
Group D: $R_D = 6(\gamma+1)/(2\gamma+1)$			
8 : [123444]	$\frac{(2\gamma+4)}{(\gamma+1)}$	$-\frac{(4\gamma-3)}{3(\gamma+1)}x$	$+\frac{((4-8t)\gamma^2+(t+1)\gamma+23t-5)}{6t(1-\gamma)(\gamma+1)}x^2$
14 : [123433]	$\frac{(2\gamma+4)}{(\gamma+1)}$	$-\frac{(2\gamma+3)}{3(\gamma+1)}x$	$+\frac{((4-8t)\gamma^2+(9t+1)\gamma+23t-5)}{6t(1-\gamma)(\gamma+1)}x^2$
20 : [123422]			
29 : [123343]			
81 : [122342]			
Group E: $R_E = 2(\gamma^2+5\gamma+3)/((\gamma+1)(2\gamma+1))$			
84 : [122333]	$\frac{(7\gamma+11)}{(\gamma^2+5\gamma+3)}$	$+\frac{(2\gamma+7)}{(\gamma^2+5\gamma+3)}x$	$+\frac{((t+7)\gamma^2+(20t-2)\gamma+21t-5)}{2t(1-\gamma)(\gamma^2+5\gamma+3)}x^2$
38 : [123322]	$\frac{(7\gamma+11)}{(\gamma^2+5\gamma+3)}$	$+\frac{(2\gamma+1)}{(\gamma^2+5\gamma+3)}x$	$+\frac{((3t+7)\gamma^2+(30t-2)\gamma+21t-5)}{2t(1-\gamma)(\gamma^2+5\gamma+3)}x^2$
85 : [122332]			
Group F: $R_F = 6/(2\gamma+1)$			
194 : [111222]	$\frac{3}{(2\gamma+1)}$	$+\frac{3}{(2\gamma+1)}x$	$+\frac{((19t-7)\gamma+17t-5)}{6t(1-\gamma)(2\gamma+1)}x^2$
104 : [122211]	$\frac{3}{(2\gamma+1)}$	$+\frac{1}{(2\gamma+1)}x$	$+\frac{((31t-7)\gamma+17t-5)}{6t(1-\gamma)(2\gamma+1)}x^2$
175 : [112221]			
Group G: $R_G = 6(\gamma+1)/(3\gamma+1)$			
33 : [123333]	$\frac{(\gamma+3)}{(\gamma+1)}$	$+\frac{(-3\gamma+4)}{3(\gamma+1)}x$	$+\frac{(-(3t-3)\gamma^2+(2t+2)\gamma+19t-5)}{6t(1-\gamma)(\gamma+1)}x^2$
55 : [123222]	$\frac{(\gamma+3)}{(\gamma+1)}$	$-\frac{2}{3(\gamma+1)}x$	$+\frac{(-(3t-3)\gamma^2+(14t+2)\gamma+19t-5)}{6t(1-\gamma)(\gamma+1)}x^2$
89 : [122322]			

Table 4.4 continues on the next page ...

4.4 Sequence Length $p = 6$

... continued from the previous page.

$l : [...]$	$h_l = c_{11}(l)$	$+2c_{12}(l)x$	$+c_{22}(l)x^2$
97 : [122232]			
Group H: $R_H = 2(5\gamma + 3)/((\gamma + 1)(3\gamma + 1))$			
174 : [112222]	$\frac{8}{(5\gamma+3)}$	$+\frac{8}{(5\gamma+3)}x$	$+\frac{((15t-3)\gamma+17t-5)}{2t(1-\gamma)(5\gamma+3)}x^2$
114 : [122111]	$\frac{8}{(5\gamma+3)}$	$+\frac{2}{(5\gamma+3)}x$	$+\frac{((27t-7)\gamma+17t-5)}{2t(1-\gamma)(5\gamma+3)}x^2$
178 : [112211]			
195 : [111221]			
Group I: $R_I = 2(2\gamma + 3)/(4\gamma + 1)$			
100 : [122222]	$\frac{5}{(2\gamma+3)}$	$+\frac{3}{(2\gamma+3)}x$	$+\frac{((3t+5)\gamma+13t-5)}{2t(1-\gamma)(2\gamma+3)}x^2$
151 : [121111]	$\frac{5}{(2\gamma+3)}$	$-\frac{3}{(2\gamma+3)}x$	$+\frac{((7t-3)\gamma+13t-5)}{2t(1-\gamma)(2\gamma+3)}x^2$
188 : [112111]			
198 : [111211]			
201 : [111121]			

Table 4.4: Some equivalence classes l, their representative sequences [...] and h_l functions for sequence length $p = 6$.

The intersection point x_2 of section 3.4, in which $h_1(x) = h_2(x)$, is located at

$$x_2 = \frac{\sqrt{RT_{x_2}} - t(1-\gamma)(\gamma + 15)}{(23t-1)\gamma^2 + (7t+1)\gamma}, \quad (4.5)$$

whereas $RT_{x_2} = t(1-\gamma)[(551t - 24)\gamma^3 + (277t + 18)\gamma^2 + 5(6 - 153t)\gamma + 225t]$ is the root term of x_2 for $p = 6$. The parameter domains are $\gamma \in (0, 1)$ and $t \geq 6$.

Equivalence class k of section 3.4 is determined as $l = 28$. The intersection point x_{28}, in which $h_1(x) = h_{28}(x)$, is located at

$$x_{28} = \frac{\sqrt{RT_{x_{28}}} - t(1-\gamma)(2\gamma^2 + 23\gamma + 27)}{2((11t-1)\gamma^3 + 30t\gamma^2 + (10t+1)\gamma - 3t)}, \quad (4.6)$$

whereas $RT_{x_{28}} = t(1-\gamma)[(524t - 48)\gamma^5 + (2672t - 120)\gamma^4 + (3799t + 24)\gamma^3 + (1171t + 120)\gamma^2 + (393t + 24)\gamma + 657t]$ is the root term of x_{28}. According to Proposition 4, the parameter domain

4 Optimal Designs

for γ is restricted to the interval $(0.3, 1)$, $t \geq 6$. Thus, the denominator of x_{28} is positive and $x_{28} \in (0, 1)$.

The 110 equivalence classes of $l \in \{2, \ldots, 202\}$ are divided into groups of identical R_u. The equivalence class l belongs to

group A $\Leftrightarrow l \in L_A = \{2, 3, 6, 7, 12, 22, 152\}$;

group B $\Leftrightarrow l \in L_B = \{9, 10, 11, 13, 15, 16, 19, 21, 25, 28, 30, 31, 36, 40, 48, 49,$
$\qquad\qquad\qquad\qquad\quad 57, 62, 70, 79, 83, 91, 105, 116, 132, 153, 157, 169\}$;

group C $\Leftrightarrow l \in L_C = \{39, 42, 52, 58, 71, 86, 170\}$;

group D $\Leftrightarrow l \in L_D = \{8, 14, 26, 32, 47, 53, 65, 74, 128, 189\}$;

group E $\Leftrightarrow l \in L_E = \{34, 35, 37, 38, 41, 43, 51, 54, 56, 59, 60, 69, 72, 73, 75, 76,$
$\qquad\qquad\qquad\qquad\quad 84, 94, 96, 98, 102, 108, 109, 112, 121, 129, 139, 143, 146,$
$\qquad\qquad\qquad\qquad\quad 158, 163, 164, 173, 190, 191, 193\}$;

group F $\Leftrightarrow l \in L_F = \{104, 111, 113, 140, 194\}$;

group G $\Leftrightarrow l \in L_G = \{33, 55, 77, 99, 131, 149, 199\}$;

group H $\Leftrightarrow l \in L_H = \{101, 103, 114, 141, 150, 174, 200\}$;

group I $\Leftrightarrow l \in L_I = \{100, 151, 202\}$.

The coefficients $c_{11}(l)$ are identical within each group, A through I. The intention is to identify the h_l function which is the maximum for each group by comparing all $c_{12}(l)$ and $c_{22}(l)$ within the group, assuming $x > 0$.

A) For $l \in L_A$, the maximal h_l function is $h_2(x)$, because $c_{12}(2) > c_{12}(l)$ and $c_{22}(2) \geq c_{22}(l)$ for all l.

B) The same criteria as in A) hold for h_{28} which is the maximum of all h_l, $l \in L_B$.

C) In group C, we have $c_{12}(170) > c_{12}(l)$ and $c_{22}(170) \geq c_{22}(l)$ for all $l \in L_C$. Thus, h_{170} is the maximum in this group.

D) The maximal h_l functions within group D are given by $l = 8$ and $l = 14$, dependent on the magnitude of γ. In this group, $c_{12}(8) > c_{12}(l)$ for all $l \in L_D$ and $c_{22}(14) \geq c_{22}(l)$ for all $l \in L_D$. The sequences l with $c_{22}(l) > c_{22}(8)$ have smaller $c_{12}(l)$ and $c_{22}(l)$ than h_{14}.

E) The same criteria as in D) hold for h_{84} and h_{38} of $l \in L_E$.

F) Again, as in D), h_{194} and h_{104} are maximal for all h_l, $l \in L_F$.

G) The maximal h_l functions within group G are given by $l = 33$ and $l = 55$, dependent on the magnitude of γ. In this group, $c_{12}(33) > c_{12}(l)$ for all $l \in L_G$ and $c_{22}(55) > c_{22}(l)$ for all $l \in L_G$. The sequences l with $c_{22}(l) > c_{22}(33)$ have smaller $c_{12}(l)$ and $c_{22}(l)$ than h_{55}.

H) The same criteria as in G) hold for h_{174} and h_{114} of $l \in L_H$.

I) Unite group I with $l = 203$. As a result, the maximal h_l is, again, represented by two equivalence classes: $l = 100$ and $l = 151$, dependent on the magnitude of γ. The result is derived from $c_{11}(100) \geq c_{11}(l)$, $c_{12}(100) > c_{12}(l)$ for all $l \in L_I \cup \{203\}$ and $c_{22}(100) > c_{22}(l)$ for $l \in L_I \cup \{203\} \setminus \{151\}$. Furthermore, $c_{22}(151) > c_{22}(l)$ for all $l \in L_I \cup \{203\}$, which yields $h_{151} > h_{100}$ by γ letting close to 1.   ✓

Lemma 10. The abscissa x_{28} is the intersection point of equivalence class functions $h_1(x) = h_{28}(x)$, cf. equation (4.6). For any $t \geq p = 6$, define the parameter $\gamma_{\beta(t)} \in (0.956, 0.958)$ as the γ-value of $t \geq 5$ for which $h_2(x_{28}) = h_{28}(x_{28})$. The maximum of all $h_l(x_{28})$, $l \in \{2,\ldots,203\}$, is $h_{28}(x_{28})$ for all $\gamma \in (\gamma_{\beta(t)}, 1)$ and all $t \geq 6$.

Proof. Consider the equivalence class functions h_l of Table 4.4 and use the results of A) through I) above. There are fifteen equivalence class functions, h_l, $l \in \{2, 28, 170, 8, 14, 84, 38, 194, 104, 33, 55, 174,$ which dominate all other h_l functions, $l \in \{2,\ldots,203\} \setminus \{2, 28, 170, 8, 14, 84, 38, 194, 104, 33, 55, 174, 114, 100, 151\}$, for $x > 0$. Since $x_{28} > 0$ for all $\gamma > 0.3$, $h_{28}(x_{28}) = \max_{l \in \{2,\ldots,203\}} h_l(x_{28})$ iff $h_{28}(x_{28}) > h_l(x_{28})$ for all $l \in \{2, 170, 8, 14, 84, 38, 194, 104, 33, 55, 174, 114, 100, 151\}$ and all $\gamma_{\beta(t)} < \gamma < 1$. For this purpose, analyze

a)
$$(h_{28} - h_2)(x) = -\frac{(4\gamma^3 + 15\gamma^2 + 18\gamma + 3)}{(\gamma+1)(\gamma+3)(2\gamma+3)} + \frac{(\gamma^3 + 11\gamma^2 + 20\gamma + 12)}{(\gamma+1)(\gamma+3)(2\gamma+3)}x$$
$$+ \frac{(7t-1)\gamma^4 + (29t-1)\gamma^3 + (41t+1)\gamma^2 + (9t+1)\gamma + 6t}{2t(1-\gamma)(\gamma+1)(\gamma+3)(2\gamma+3)}x^2.$$

Substitution of $x = x_{28}$ yields
$$(h_{28} - h_2)(x_{28}) = \frac{b(\gamma)\sqrt{RT_{x_{28}}} - (1-\gamma)a(\gamma)}{4(2\gamma+3)[(11t-1)\gamma^3 + 30t\gamma^2 + (10t+1)\gamma - 3t]^2},$$

with
$$a(\gamma) = -8(9t^2 + 10t - 1)\gamma^6 + 2t(23t - 117)\gamma^5 + (1745t^2 - 71t - 8)\gamma^4 + 45t(41t+1)\gamma^3 - t(1147t + 29)\gamma^2 - 3t(93t+5)\gamma + 774t^2$$

and
$$b(\gamma) = 8t\gamma^4 + 3(17t+1)\gamma^3 - 2(23t+1)\gamma^2 - (35t+1)\gamma + 30t.$$

The denominator of $(h_{28} - h_2)(x_{28})$ is positive for all γ and t in the corresponding domains. In order to determine the roots of $b(\gamma)\sqrt{RT_{x_{28}}} - (1-\gamma)a(\gamma)$, the algebraic signs of $a(\gamma)$ and $b(\gamma)$ need to be checked.

4 Optimal Designs

a1) Abbreviate $a(\gamma) = a_0 + a_1\gamma + \ldots + a_6\gamma^6$. The coefficients a_1, a_2 and a_6 are negative. We have $a_6\gamma^6 + a_5\gamma^5 + a_4\gamma^4 > (a_6 + a_5 + a_4)\gamma^4 = (1719t - 385)t\gamma^4 > 0$ for all γ and all $t \geq 6$.

Now, define the cubic function of γ, $A_{03}(\gamma) := a_0 + a_1\gamma + a_2\gamma^2 + a_3\gamma^3$. The second derivative of $A_{03}(\gamma)$ equals 0 iff $\gamma = \gamma_{inf} := \frac{1147t+29}{135(41t+1)}$. $A_{03}(\gamma)$ is concave for all $\gamma < \gamma_{inf}$ and convex for all $\gamma > \gamma_{inf}$.

There is one stationary point of $A_{03}(\gamma)$ in $(0,1)$ located at $\gamma_{st} = \frac{\sqrt{2(1429937t^2+93608t+1433)}+1147t+29}{135(41t+1)}$. The convexity of $A_{03}(\gamma)$ implies that $A_{03}(\gamma_{st})$ is a local minimum of $A_{03}(\gamma)$, since $\gamma_{st} > \gamma_{inf}$. Some simple calculus confirms that

$$A_{03}(\gamma_{extr}) = \frac{t(V - 4\sqrt{2W^3})}{54675(41t+1)^2}$$

is positive for all $t \geq 6$. The values of V and W are given as $V = 62805491539t^3 + 2691551589t^2 + 19062633t - 224953$ and $W = 1429937t^2 + 93608t + 1433$ in the formula of $A_{03}(\gamma_{extr})$. As $A_{03}(\gamma_{extr})$ is positive, the function $A_{03}(\gamma)$ is positive for all $\gamma \in (0,1)$ and all $t \geq 6$.

To summarize, the function $a(\gamma) = A_{03}(\gamma) + a_4\gamma^4 + a_5\gamma^5 + a_6\gamma^6$ is positive for all $\gamma \in (0,1)$ and all $t \geq 6$.

a2) In order to prove that $b(\gamma)$ is positive for all $\gamma \in (0,1)$, examine the derivative of the function.

The function $b(\gamma)$ has one point of inflection in the domain $(0,1)$ of γ at $\gamma_{inf} = \frac{1}{96t}[\sqrt{10747t^2 + 1046t + 27} - 3(17t+1)]$. Therefore, $b(\gamma)$ is concave for all $\gamma < \gamma_{inf}$ and convex, otherwise. The local minimum of the concave part of $b(\gamma)$ is either located at 0 or γ_{inf}. The function value of $b(\gamma \searrow 0) = b_0$ is positive. Simple calculus provides that

$b(\gamma_{inf}) = [3(189803t^3 + 30577t^2 + 1569t + 27)\sqrt{3(10747t^2 + 1046t + 27)} - (99598873t^4 + 21481252t^3 + 1654486t^2 + 56484t + 729)]/147456t^3$.

is positive as well. Thus, $b(\gamma) > 0$ for all $\gamma \in (0, \gamma_{inf})$. Equivalence transformations confirm that $\gamma_{inf} > 0.2$.

In order to analyze $b(\gamma)$ in the remaining interval of $(\gamma_{inf}, 1)$, apply the transformation $\gamma = 0.2 + 0.8\gamma'$ with $\gamma' \in (0,1)$ and $\gamma \in (0.2, 1)$. This results in

$b(\gamma') = [2048t\gamma'^4 + 64(287t+15)\gamma'^3 - 16(337t+5)\gamma'^2 - 8(2939t+90)\gamma' + 16(843t-10)]/625$.

4.4 Sequence Length $p = 6$

Abbreviate $b(\gamma')$ as $(b_0 + b_1\gamma' + \ldots + b_4\gamma'^4)/625$ and observe that the coefficients b_0, b_3 and b_4 are positive, b_2 and b_1 are negative. In order to verify whether $b(\gamma') > 0$, define $B_{03}(\gamma') := b(\gamma') - b_4\gamma'^4$. There is one stationary point of $B_{03}(\gamma')$ in $(0,1)$ at $\gamma_{st} = \frac{\sqrt{5174527t^2+422860t+8125}+337t+5}{12(287t+15)}$. A point of inflection of $B_{03}(\gamma')$ is located at $\gamma_{inf} = \frac{337t+5}{12(287t+15)}$, such that $B_{03}(\gamma')$ is concave for all $\gamma' < \gamma_{inf}$ because $B_{03}''(\gamma' < \gamma_{inf}) < 0$, and convex for all $\gamma' > \gamma_{inf}$. It is easy to derive that $\gamma_{inf} < \gamma_{st}$. Hence, $B_{03}(\gamma_{st})$ yields the minimum of $B_{03}(\gamma')$.

$$B_{03}(\gamma_{st}) = \frac{2(25V - \sqrt{W^3})}{27(287t+15)^2},$$

in which $V = 496072778t^3 + 45525909t^2 + 604248t - 21875$ and $W = 5174527t^2 + 422860t + 8125$. Some simple equivalence transformations confirm that $B_{03}(\gamma_{st})$ is positive for all $\gamma' \in (0,1)$ and all $t \geq 6$.

As $b(\gamma') = B_{03}(\gamma') + b_4\gamma'^4 > 0$ for all $\gamma' \in (0,1)$, $b(\gamma)$ is positive for all $\gamma \in [0.2, 1) \cup (0, 0.2)$ and all $t \geq 6$.

The results of a1) and a2) lead to an equivalent transformation of $b(\gamma)\sqrt{RT_{x28}} - (1-\gamma)a(\gamma) > 0$, in which $(1-\gamma)a(\gamma)$ is added and the square is taken on both sides of the inequality. This is equivalent to

$$b^2(\gamma)RT_{x28} - (1-\gamma)^2 a^2(\gamma) > 0$$
$$\Leftrightarrow 8(1-\gamma)((11t-1)\gamma^3 + 30t\gamma^2 + (10t+1)\gamma - 3t)^2 \cdot g(\gamma) > 0$$
$$\Leftrightarrow g(\gamma) := g_0 + g_1\gamma + \ldots + g_7\gamma^7 > 0,$$

in which

$g(\gamma) = 8(5t^2+2t+1)\gamma^7 + 4(97t^2-t-2)\gamma^6 + 8t(97t-4)\gamma^5 - t(2267t+43)\gamma^4 + 3t(135t+11)\gamma^3 + 9t(161t+3)\gamma^2 - 3t(217t-1)\gamma - 108t^2.$

In order to find the roots of $g(\gamma)$, decompose $g(\gamma)$ into $G_{-4}(\gamma) = g(\gamma) - g_4\gamma^4$ and $g_4\gamma^4$. Since g_4 is negative, g_4 is strictly monotonous and decreasing in t for all $\gamma \in (0,1)$ and $t \geq 6$.

All coefficients g_i, $i \in \{2,3,5,6,7\}$, are positive. Thus, the curvature, determined by the second derivative of $G_{-4}(\gamma)$, is positive for all $\gamma \in (0,1)$, and the function $G_{-4}(\gamma)$ is convex in the entire domain of γ. The convexity of $G_{-4}(\gamma)$ and different signs of the function values for the range boundaries, $G_{-4}(\gamma \searrow 0) = g_0 < 0$ and $G_{-4}(\gamma \nearrow 1) = (1095t + 63)t > 0$, imply that there can only exist one root of $G_{-4}(\gamma)$ in the interval $(0,1)$.

4 Optimal Designs

It is clear that $g(\gamma)$ is negative iff $G_{-4}(\gamma)$ is negative because $g_4 < 0$.
In the range of $G_{-4}(\gamma) > 0$, $G_{-4}(\gamma)$ is a monotonous and increasing function of γ. The function $g(\gamma)$ is positive iff $|g_4\gamma^4| < G_{-4}(\gamma)$, assuming $G_{-4}(\gamma) > 0$. Iff $|g_4\gamma^4| > G_{-4}(\gamma)$, the function $g(\gamma)$ is negative again.
As $g_4 < 0$ is strictly monotonous and $G_{-4}(\gamma)$ is monotonous in the interval in which $G_{-4}(\gamma) > 0$, it follows that the sum $g(\gamma)$ of those two monotonous functions can only change its algebraic signs once, most likely at the range boundary $\gamma \nearrow 1$. The function value of this range boundary is $g(\gamma \nearrow 1) = 32t^2 > 0$ for all $t \geq 6$. Therefore, $g(\gamma)$ actually can only have one root $\gamma_{\beta(t)}$, which is located in the interval $(0.956, 0.958)$. The root $\gamma_{\beta(t)}$ is the specific parameter at which $h_2(x_{28}) = h_{28}(x_{28})$ for $t \geq 5$. Thus, $g(\gamma)$ is positive for all $\gamma > \gamma_{\beta(t)}$.
Hence, $(h_{28} - h_2)(x_{28}) > 0$ iff $\gamma > \gamma_{\beta(t)}$ which leads to $h_{28}(x_{28}) > h_l(x_{28})$ for $l \in L_A$ and $\gamma \in (\gamma_{\beta(t)}, 1)$.

b) through o) are treated in Appendix B, section B.2.2

Finally, as $x_{28} \in (0,1)$, $h_{28}(x_{28}) = \max_l h_l(x_{28})$ for all $2 \leq l \leq 203$ iff $\gamma \in (\gamma_{\beta(t)}, 1)$. $\square$

Lemma 11. The abscissa x_2 is an intersection point of equivalence class functions $h_1(x) = h_2(x)$, cf. equation (4.5). Analogous to Lemma 10, consider the same parameter $\gamma_{\beta(t)} \in (0.956, 0.958)$ as the γ-value for which $h_2(x_{28}) = h_{28}(x_{28})$ and $h_2(x_2) = h_{28}(x_2)$ as well, i.e., $x_2 = x_{28}$, for any $t \geq p = 6$.
The maximum of $h_l(x_2)$, $l \in \{2, \ldots, 203\}$, is $h_2(x_2)$ for all $\gamma \in (0, \gamma_{\beta(t)})$ and all $t \geq 6$.

Proof. Refer to the equivalence class functions h_l of Table 4.4 and use the results of A) through I) above. There are fifteen equivalence class functions h_l, $l \in \{2, 28, 170, 8, 14, 84, 38, 194, 104, 33, 55, 174, 114, 100, 151\}$, which dominate all other h_l functions, $l \in \{2, \ldots, 203\} \setminus \{2, 28, 170, 8, 14, 84, 38, 194, 104, 33, 55, 174, 114, 100, 151\}$, for $x > 0$. Since $x_2 > 0$, $h_2(x_2) = \max_{l \in \{2,\ldots,203\}} h_l(x_2)$ iff $h_2(x_2) > h_l(x_2)$ for all $l \in \{28, 170, 8, 14, 84, 38, 194, 104, 33, 55, 174, 114, 100, 151\}$ and all $0 < \gamma < \gamma_{\beta(t)}$. We have to analyze

a)
$$(h_2 - h_{28})(x) = \frac{(4\gamma^3 + 15\gamma^2 + 18\gamma + 3)}{(\gamma+1)(\gamma+3)(2\gamma+3)} - \frac{(\gamma^3 + 11\gamma^2 + 20\gamma + 12)}{(\gamma+1)(\gamma+3)(2\gamma+3)} x$$
$$+ \frac{-(7t-1)\gamma^4 - (29t-1)\gamma^3 - (41t+1)\gamma^2 - (9t+1)\gamma + 6t}{2t(1-\gamma)(\gamma+1)(\gamma+3)(2\gamma+3)} x^2.$$

Substitution of $x = x_2$ provides

$$(h_2 - h_{28})(x_2) = \frac{(1-\gamma)a(\gamma) - b(\gamma)\sqrt{RT_{x_2}}}{\gamma^2(1+\gamma)(\gamma+3)[(23t-1)\gamma + 7t + 1]^2},$$

with

$a(\gamma) = -4(21t^2 + 22t - 1)\gamma^5 + (143t^2 - 29t - 4)\gamma^4 + (1271t + 19)t\gamma^3 - t(419t + 13)\gamma^2 - 3t(151t + 3)\gamma + 450t^2$

and

$b(\gamma) = 8t\gamma^4 + 3(17t+1)\gamma^3 - 2(23t+1)\gamma^2 - (35t+1)\gamma + 30t.$

The denominator of $(h_2 - h_{28})(x_2)$ is positive for all γ and t in the stated domains. In order to determine the root of $(h_2 - h_{28})(x_2)$ to be located at $\gamma = \gamma_{\beta(t)}$, it is convenient to take advantage of $b(\gamma)$ being identical to $b(\gamma)$ of $(h_{28} - h_2)(x_{28})$ in the proof of Lemma 10 a). Thus, $b(\gamma)$ is positive for all $\gamma \in (0,1)$ and all $t \geq 6$.

Function $a(\gamma)$ is positive for all $\gamma \in (0,1)$ and all $t \geq 6$ as well. Define $a(\gamma) = A_{45}(\gamma) + A_{03}(\gamma)$ and rewrite $A_{45}(\gamma) = a_5\gamma^5 + a_4\gamma^4$ and $A_{03}(\gamma) = a_3\gamma^3 + a_2\gamma^2 + a_1\gamma + a_0$. Use the fact that $a_5\gamma^5 > a_5\gamma^4$ to get $A_{45}(\gamma) > (a_5 + a_4)\gamma^4 = t(59t - 117)\gamma^4 > 0$ for all $\gamma \in (0,1)$ and all $t \geq 6$.

Furthermore, the second derivative of $A_{03}(\gamma)$ implies that $A_{03}(\gamma)$ is concave for all $\gamma < \gamma_{inf}$, and convex for all $\gamma > \gamma_{inf}$, in which $\gamma_{inf} = \frac{419t+13}{3(1271t+19)}$. The slope of $A_{03}(\gamma)$ is zero in interval $(0,1)$ iff $\gamma = \gamma_{st} = \frac{\sqrt{2(951425t^2 + 35516t + 341)} + 419t + 13}{3813t + 57}$. Compare γ_{inf} and γ_{st} to find that $\gamma_{inf} < \gamma_{st}$. Thus, $A_{03}(\gamma_{st})$ yields the minimum of $A_{03}(\gamma)$. Some calculus ensures that

$$A_{03}(\gamma_{st}) = \frac{t(V - 4\sqrt{2W^3})}{(43616907t^2 + 1304046t + 9747)}$$

is positive for all $t \geq 6$. The substitutes in the formula of $A_{03}(\gamma_{st})$ are given as $V = 17309285759t^3 + 430169205t^2 + 971061t - 24401$ and $W = 951425t^2 + 35516t + 341$. It follows that $A_{03}(\gamma)$ is positive for all $\gamma \in (0,1)$ and all $t \geq 6$. As $a(\gamma) = A_{45}(\gamma) + A_{03}(\gamma)$, $a(\gamma)$ is positive for all $\gamma \in (0,1)$ and all $t \geq 6$ as well.

Therefore, $(h_2 - h_{28})(x_2) > 0$ is equivalent to

$$(1-\gamma)^2 a^2(\gamma) - b^2(\gamma) RT_{x_2} > 0$$
$$\Leftrightarrow 2(1-\gamma)\gamma^2[(23t-1)\gamma + 7t + 1]^2 \cdot g^*(\gamma) > 0$$
$$\Leftrightarrow g^*(\gamma) > 0,$$

in which $g^*(\gamma) = -g(\gamma)$ of the difference $(h_{28} - h_2)(x_{28})$, cf. passage a) of proof of Lemma 10. This symmetry implies that $g^*(\gamma)$ and $g(\gamma)$ have the same root for all $\gamma \in (0,1)$ at

4 Optimal Designs

$\gamma = \gamma_{\beta(t)} \in (0.956, 0.958)$ and $\gamma_{\beta(t)}$ is the specific parameter in which $x_2 = x_{28}$. Hence, $h_2(x_2) > h_{28}(x_2)$ for all $\gamma \in (0, \gamma_{\beta(t)})$.

b) Now, compare the equivalence class functions $h_l(x)$, $l \in \{8, 14, 194, 104, 33, 174, 100, 151\}$, to $h_2(x)$. It is known, so far, that $h_{28}(x) > h_l(x)$ for $l \in \{8, 14, 194, 104, 33, 174, 100, 151\}$ and all $x, \gamma \in (0,1)$, cf. proof of Lemma 10, statements c), d), g), h), i), l), n) and o); $x_2 \in (0,1)$, cf. Proposition 3; and $h_2(x_2) > h_{28}(x_2)$ for all $\gamma \in (0, \gamma_{\beta(t)}) \subset (0,1)$. Combining the three results, it follows that $h_2(x_2) > h_l(x_2)$ for all $l \in \{L_D, L_F, \{33, 174, 203\}, L_I\}$ and all $\gamma \in (0, \gamma_{\beta(t)})$.

It remains to show whether $h_2(x_2) > h_l(x_2)$ for all $l \in \{170, 84, 38, 55, 114\}$ and all $\gamma \in (0, \gamma_\beta)$.

c) through g) continues in Appendix B, section B.2.2

Summarizing statements a) through g), the conclusion is that $h_2(x_2) = \max_l h_l(x_2)$ for all $l \in \{L_A, L_B, L_C, L_D, L_E, L_F, L_G, L_H, L_I, 203\}$ and, thus, for all $2 \le l \le 203$, and all $\gamma \in (0, \gamma_\beta)$. □

Lemma 12. For any $t \ge p = 6$, the parameters $0 < \gamma_{\alpha 1} < \gamma_{\alpha 2} < 1$ are given by

$$\gamma_{\alpha 1 / \alpha 2} = \frac{\mp\sqrt{339889 t^4 - 317254 t^3 + 108061 t^2 - 16140 t + 900} + 583 t^2 - 209 t + 18}{2(613 t^2 - 239 t + 24)}.$$

Assume $\gamma \in (\gamma_{\alpha 1}, \gamma_{\alpha 2})$ and observe that $x_{min} = t/(5t - 1)$ is the abscissa of the minimum of equivalence class function $h_1(x)$. Then, $h_1(x_{min}) = \max_{l \in \{1,\ldots,203\}} h_l(x_{min})$.

Proof. Since $x_{min} > 0$, we may use the results of A) through I) above, i.e., one of the h_l, $l \in \{2, 28, 170, 8, 14, 84, 38, 194, 104, 33, 55, 174, 114, 100, 151\}$ is the maximum of all other h_l functions, $l \in \{2, \ldots, 203\} \setminus \{2, 28, 170, 8, 14, 84, 38, 194, 104, 33, 55, 174, 114, 100, 151\}$, for any $x > 0$. Using the results of the proof of Lemma 10, i.e., $h_{28} = \max_{l \in \{8, 14, 194, 104, 33, 174, 100, 151\}} h_l$ for all $x \in (0,1)$, it remains to verify whether $h_1(x_{min}) > h_l(x_{min})$ for $l \in \{2, 28, 170, 84, 38, 55, 114\}$. It follows that $h_1(x_{min}) = \max_{l=1,\ldots,203} h_l(x_{min})$. Now we have

a)
$$(h_2 - h_1)(x_{min}) = \frac{(613 t^2 - 239 t + 24)\gamma^2 - (583 t^2 - 209 t + 18)\gamma + 6(5t - 1)}{6(5t - 1)^2 (1 - \gamma)(2\gamma + 3)}.$$

The solution of the quadratic equation $(613 t^2 - 239 t + 24)\gamma^2 - (583 t^2 - 209 t + 18)\gamma + 6(5t - 1) = 0$ for all $\gamma \in (0,1)$ and all $t \ge 6$ indicates

$$(h_2 - h_1)(x_{min}) \begin{cases} < 0 & \Leftrightarrow \gamma \in (\gamma_{\alpha 1}, \gamma_{\alpha 2}) \\ \ge 0 & \Leftrightarrow \gamma \notin (\gamma_{\alpha 1}, \gamma_{\alpha 2}) \end{cases}.$$

Hence, $h_1(x_{min}) > h_2(x_{min})$ for $\gamma \in (\gamma_{\alpha 1}, \gamma_{\alpha 2})$.

b) through g) is treated in Appendix B, section B.2.2.

Combining properties a) through g) shows that $h_1(x_{min})$ is the maximum of the $h_l(x_{min})$ for all $\gamma \in (\gamma_{\alpha 1}, \gamma_{\alpha 2})$ and all $l \in \{2, 28, 170, 84, 38, 55, 114\}$, and thus, for all $2 \leq l \leq 203$. □

Theorem 4. For any $t \geq p = 6$ and $\gamma \in (0, 1)$, consider the parameters $0 < \gamma_{\alpha_1} < \gamma_{\alpha_2} < \gamma_{\beta(t)} < 1$ as in Lemmas 12 and 11. Furthermore, proportion $\alpha(\gamma) \in [0, 1]$ is given by

$$\alpha(\gamma) = \frac{5(1-\gamma)(2\gamma+3)[3t(6t\gamma^2 - (21t-5)\gamma + 5(5t-1)) - (5t-1)\sqrt{RT_{x_2}}]}{\gamma[(23t-1)\gamma + 7t + 1]\sqrt{RT_{x_2}}}$$

iff $\gamma \in (0, \gamma_{\alpha 1}] \overset{\bullet}{\cup} [\gamma_{\alpha 2}, \gamma_{\beta(t)}]$ and $\alpha(\gamma) := 0$ iff $\gamma \in (\gamma_{\alpha 1}, \gamma_{\alpha 2})$ or $\gamma \in (\gamma_{\beta(t)}, 1)$. Another proportion $\beta(\gamma) \in [0, 1]$ is given by

$$\beta(\gamma) = \frac{5(1-\gamma^2)(\gamma+3)[3t(4t\gamma^3 + (-15t+7)\gamma^2 + 2\gamma + 43t - 9) - (5t-1)\sqrt{RT_{x_{28}}}]}{2[(11t-1)\gamma^3 + 30t\gamma^2 + (10t+1)\gamma - 3t]\sqrt{RT_{x_{28}}}}$$

iff $\gamma \in [\gamma_{\beta(t)}, 1)$ and $\beta(\gamma) := 0$ otherwise.
The optimality results are as follows:

a) For all $\gamma \in (0, \gamma_{\beta(t)})$, an approximate design d^* is optimal iff $(1 - \alpha(\gamma)) \cdot 100\%$ of its sequences are selected from class 1 with representative sequence $[1, 2, 3, 4, 5, 6]$ and $\alpha(\gamma) \cdot 100\%$ of its sequences from class 2 with representative sequence $[1, 2, 3, 4, 5, 5]$.

b) For all $\gamma \in (\gamma_{\beta(t)}, 1)$, an approximate design d^* is optimal iff $(1 - \beta(\gamma)) \cdot 100\%$ of its sequences are selected from class 1 and $\beta(\gamma) \cdot 100\%$ of its sequences from class 28 with representative sequence $[1, 2, 3, 3, 4, 4]$.

c) Iff $\gamma = \gamma_{\beta(t)}$, define proportion $\varphi \in [0, 1]$. An approximate design d^* is optimal iff $\varphi \cdot 100\%$ of the sequences are arranged as in a) and $(1 - \varphi) \cdot 100\%$ of the sequences are arranged as in b).

Proof. Theorem 4 indicates that x_{d^*} of Proposition 2 in which $\min_{x} \max_{l} h_l(x)$ is being realized, is either x_2, x_{28}, $x_2 = x_{28}$, or x_{min}, the x-coordinate of the minimum of h_1. In order to verify this conjecture, eight properties and formulas need to be derived:

1. For all $\gamma \notin (\gamma_{\alpha 1}, \gamma_{\alpha 2})$ and $\gamma < \gamma_{\beta(t)}$: $h_2(x_2) > h_l(x_2)$ for all $3 \leq l \leq 203$.

2. For all $\gamma \notin (\gamma_{\alpha 1}, \gamma_{\alpha 2})$: sign $h'_1(x_2) \neq$ sign $h'_2(x_2)$.

4 Optimal Designs

3. For all $\gamma \in (\gamma_{\alpha 1}, \gamma_{\alpha 2})$: $h_1(x_{min}) > h_l(x_{min})$ for all $2 \leq l \leq 203$.

4. The formula of $\alpha(\gamma)$.

5. For all $\gamma \in (\gamma_{\beta(t)}, 1)$: $h_{28}(x_{28}) > h_l(x_{28})$ for all $2 \leq l \leq 27$ and $29 \leq l \leq 203$.

6. For all $\gamma \in (\gamma_{\beta(t)}, 1)$: $\text{sign } h'_1(x_{28}) \neq \text{sign } h'_{28}(x_{28})$.

7. The formula of $\beta(\gamma)$.

8. For $\gamma = \gamma_{\beta(t)}$: $x^* := x_2 = x_{28}$ and $h_2(x^*) = h_{28}(x^*) > h_l(x^*)$ for all $3 \leq l \leq 27$ and $29 \leq l \leq 203$.

Properties 1, 3 and 5 follow from Lemmas 11, 12 and 10, respectively. Statement 8 is proved by Lemmas 10 and 11. ✓

As required in point 4, the proportion $\alpha(\gamma) \in (0,1)$ of equivalence class 2 sequences needs to be determined. For this purpose, use equation (2.8) and write $\alpha(\gamma, x) = \alpha$. Then $q_{d^*}(x) = \alpha n h_2(x) + (1-\alpha)n h_1(x)$. The proportion $\alpha(\gamma, x)$ results from setting

$$0 \stackrel{!}{=} \frac{\partial q_{d^*}}{\partial x} = \alpha \left(\frac{5}{3} - \frac{3\gamma}{2\gamma + 3} + \frac{(1-3t)\gamma(2+3\gamma) + 5(5t-1)}{t(1-\gamma)(2\gamma+3)} x - \frac{5(5t-1)}{3t} x \right)$$
$$- \frac{5}{3} + \frac{5(5t-1)}{3t} x$$
$$\Leftrightarrow \alpha(\gamma, x) = \frac{5[(1-5t)x + t](1-\gamma)(2\gamma+3)}{x\gamma[(23t-1)\gamma + 6t + 1] + t(1-\gamma)(\gamma+15)}.$$

Substitution of $x = x_2$ into the formula of $\alpha(\gamma, x)$ provides $\alpha(\gamma)$. A graphical presentation of $\alpha(\gamma)$ for $t = p = 6$ in comparison with $t = p = 5$ is displayed in Figure 4.4. A plot of $\alpha(\gamma)$ for $p = 6$ and different t is qualitatively equivalent to Figure 4.3. It is important to consider the downsizing in the scale of the proportions and the shifting of $\gamma_{\alpha 1}$ approaching 0, and $\gamma_{\alpha 2} < \gamma_{\beta(t)}$ both getting closer to 1 for increasing t. ✓

The condition to achieve that $\alpha(\gamma) \in (0,1)$ is $\text{sign } h'_1(x_2) \neq \text{sign } h'_2(x_2)$. Proportion $\alpha(\gamma)$ is negative or exceeds 1 iff $\text{sign } h'_1(x_2) = \text{sign } h'_2(x_2)$. However, this is not valid as an equivalence class proportion. In order to prove statement 2, it is sufficient to analyze if $\alpha(\gamma)$, or $\alpha(\gamma, x)$ respectively, is nonnegative and less or equal to 1 in the described domains of the parameters

t and γ. Thus, observe

$$\alpha(\gamma, x = x_2) \stackrel{!}{=} 0 \Leftrightarrow (1 - 5t)x_2 + t = 0$$
$$\stackrel{(4.5)}{\Leftrightarrow} t\gamma[(23t - 1)\gamma + 7t + 1]$$
$$\cdot [(613t^2 - 239t + 24)\gamma^2 - (583t^2 - 209t + 18)\gamma + 6(5t - 1)] = 0$$
$$\Leftrightarrow \gamma \in \{\gamma_{\alpha 1}, \gamma_{\alpha 2}, -(7t + 1)/(23t - 1), 0\}.$$

The last two elements $-(7t+1)/(23t-1)$ and 0 are not in the required domain of γ. As $t\gamma[(23t-1)\gamma + 7t + 1] > 0$ for all $\gamma > 0$ and $t > 0$, and $[(613t^2 - 239t + 24)\gamma^2 - (583t^2 - 209t + 18)\gamma + 6(5t - 1)]$ is a convex parabola in γ, proportion $\alpha(\gamma) \geq 0$ iff $\gamma \notin (\gamma_{\alpha 1}, \gamma_{\alpha 2})$.

Next, the proportion $\alpha(\gamma)$ needs to be less or equal to 1, which is fulfilled iff the numerator, $A - B\sqrt{RT_{x_2}}$, of $\alpha(\gamma)$ is less or equal to the denominator, $C\sqrt{RT_{x_2}}$, of $\alpha(\gamma)$ in which

$A = \quad 5(1 - \gamma)(2\gamma + 3) \cdot 3t[6t\gamma^2 - (21t - 5)\gamma + 5(5t - 1)],$
$B = \quad 5(1 - \gamma)(2\gamma + 3)(5t - 1)$ and
$C = \quad \gamma[(23t - 1)\gamma + 7t + 1].$

All expressions A, B, C are positive as $\gamma \in (0, 1)$ and $t \geq 5$, i.e.,

$$A - B\sqrt{RT_{x_2}} \leq C\sqrt{RT_{x_2}}$$
$$\Leftrightarrow A^2 - RT_{x_2}(B + C)^2 \leq 0$$
$$\Leftrightarrow 27t(1 - \gamma)\gamma[(23t - 1)\gamma + 7t + 1]f(\gamma) \leq 0$$
$$\Leftrightarrow f(\gamma) \leq 0,$$

for $\gamma \in (0, \gamma_{\beta(t)})$, whereas

$f(\gamma) = \quad -3(233t^2 - 143t + 24)\gamma^5 - 2(357t^2 - 295t + 57)\gamma^4 + (2965t^2 - 1537t + 184)\gamma^3 + (2348t^2 - 1512t + 212)\gamma^2 - (3650t^2 - 1530t + 160)\gamma - (1250t^2 - 500t + 50).$

The value $\gamma_{\beta(t)}$ is the root of function

$g(\gamma) = \quad 8(5t^2 + 2t + 1)\gamma^7 + 4(97t^2 - t - 2)\gamma^6 + 8t(97t - 4)\gamma^5 - t(2267t + 43)\gamma^4 + 3t(135t + 11)\gamma^3 + 9t(161t + 3)\gamma^2 - 3t(217t - 1)\gamma - 108t^2$

and simultaneously the value of $\gamma \in (0, 1)$ at which $x_2 = x_{28}$, cf. Lemmas 10 and 11. As already displayed in property a2) of proof of Lemma 10, $g(\gamma)$ is negative for all $\gamma \in (0, \gamma_{\beta(t)})$. Hence, if $f(\gamma) \leq g(\gamma)$ for all $\gamma \in (0, \gamma_{\beta(t)})$, $f(\gamma)$ is negative for this described interval of γ and $\alpha(\gamma)$ would be less or equal to 1.

Define $F(\gamma) = f(\gamma) - g(\gamma) = f_7\gamma^7 + \ldots + f_1\gamma + f_0$, i.e.,

4 Optimal Designs

$$F(\gamma) = -8(5t^2 + 2t + 1)\gamma^7 - 4(97t^2 - t - 2)\gamma^6 - (1475t^2 - 461t + 72)\gamma^5 + \\ (1553t^2 + 633t - 114)\gamma^4 + 2(1280t^2 - 785t + 92)\gamma^3 + (899t^2 - 1539t + \\ 212)\gamma^2 - (2999t^2 - 1527t + 160)\gamma - 2(571t^2 - 250t + 25).$$

The coefficients f_7, f_6, f_1 and f_0 are negative, f_5, f_4, f_3 and f_2 are positive. Thus, if $F_{05}(\gamma) := f_5\gamma^5 + \ldots + f_1\gamma + f_0$ is negative, the entire function $F(\gamma)$ is negative.

Now, analyze the second derivative of $F_{05}(\gamma)$. We have $F_{05}''(\gamma) = 20f_5\gamma^3 + 12f_4\gamma^2 + 12f_3\gamma + 2f_2 > \underbrace{(20f_5 + 12f_4)}_{<0}\gamma^2 + 12f_3\gamma + 2f_2 > \underbrace{(20f_5 + 12f_4 + 12f_3)}_{>0}\gamma + 2f_2 > 0$, i.e., $F_{05}(\gamma)$ is a convex function and its local maxima is either the left or right range boundary of interval $(0, \gamma_{\beta(t)}) \subset (0, 0.958)$. Observe $F_{05}(\gamma \searrow 0) = f_0 < 0$ and $F_{05}(0.958) = -\frac{27857412064829501t^2 - 2551129495924573t + 35849766755040}{33918856765500} < 0$ for all $t \geq 6$. Hence, $F_{05}(\gamma)$ is negative for all $\gamma \in (0, 0.958) \supset (0, \gamma_{\beta(t)})$. Thus, $F(\gamma) = f(\gamma) - g(\gamma) < 0 \Leftrightarrow f(\gamma) < g(\gamma) < 0$ for all $\gamma \in (0, \gamma_{\beta(t)})$. It follows that $\alpha(\gamma) \leq 1$ for all $\gamma \in (0, \gamma_{\beta(t)})$. Hence, sign $h_1'(x_2) \neq$ sign $h_2'(x_2)$ for all $\gamma \notin (\gamma_{\alpha 1}, \gamma_{\alpha 2})$. ✓

The proportion $\beta(\gamma) \in (0, 1)$ of equivalence class 28 sequences needs to be determined in statement 7. Therefore, use equation (2.8) and write $\beta(\gamma, x) = \beta$. Then, $q_{d^*}(x) = \beta n h_{28}(x) + (1 - \beta) n h_1(x)$. The proportion $\beta(\gamma, x)$ is calculated from setting

$$0 \stackrel{!}{=} \frac{\partial q_{d^*}}{\partial x} = \\ = \beta\left(\frac{5}{3} + \frac{4 + \gamma - \gamma^2}{(\gamma + 1)(\gamma + 3)} + \frac{(1-t)\gamma^3 + 5(1-t)\gamma^2 + (15t+1)\gamma + 23t - 5]}{t(1-\gamma^2)(\gamma + 3)}x\right.\\ \left. - \frac{5(5t-1)}{3t}x\right) - \frac{5}{3} + \frac{5(5t-1)}{3t}x$$

$$\Leftrightarrow \beta(\gamma, x) = \frac{5[(1-5t)x + t](1-\gamma^2)(\gamma + 3)}{2[(11t-1)\gamma^3 + 30t\gamma^2 + (10t+1)\gamma - 3t]x + t(1-\gamma)(2\gamma^2 + 23\gamma + 27)}.$$

Substitution of $x = x_{28}$ into the formula of $\beta(\gamma, x)$ provides $\beta(\gamma)$. A graphical presentation of $\beta(\gamma)$ for $t = p = 6$ in comparison to $t = p = 5$ is displayed in Figure 4.4. For $p = 6$ and different t, a plot of $\beta(\gamma)$, as well as $\alpha(\gamma)$, is qualitatively equivalent to Figure 4.3. ✓

Behaving like $\alpha(\gamma)$, proportion $\beta(\gamma)$ is negative or exceeds 1 iff sign $h_1'(x_{28}) = $ sign $h_{28}'(x_{28})$. In order to prove statement 6, it is sufficient to analyze if $\beta(\gamma)$, or $\beta(\gamma, x)$ respectively, is nonnegative and less or equal to 1 for $t \geq 6$ and $\gamma \in (\gamma_{\beta(t)}, 1)$. For this purpose, observe that

$$\beta(\gamma, x = x_{28}) \stackrel{!}{\geq} 0 \Leftrightarrow (1 - 5t)x_{28} + t \geq 0 \stackrel{(4.6)}{\Leftrightarrow} \frac{4t \cdot a(\gamma) \cdot b(\gamma)}{(5t - 1)^2} > 0,$$

4.4 Sequence Length $p = 6$

whereas

$$a(\gamma) = (11t-1)\gamma^3 + 30t\gamma^2 + (10t+1)\gamma - 3t$$

and

$$b(\gamma) = (301t^2 - 119t + 12)\gamma^3 + 3(125t^2 - 53t + 6)\gamma^2 - (610t^2 - 245t + 24)\gamma - 3(6t^2 - 11t + 2).$$

The proportion $\beta(\gamma)$ is positive iff $a(\gamma)$ and $b(\gamma)$ are both positive or both negative for all $\gamma \in (\gamma_{\beta(t)}, 1)$.

The function $a(\gamma)$ is equal to a factor of the denominator of x_{28}, which is positive for all $\gamma \in (0.3, 1)$, cf. equation (4.6) and Proposition 4. As $(\gamma_{\beta(t)}, 1)$ is a subset of $(0.952, 1)$, apply the transformation $\gamma = 0.048\gamma' + 0.952$ to $b(\gamma)$, in which $\gamma' \in (0,1)$ and $\gamma \in (0.92, 1)$. This transformation gives

$$b(\gamma') = 6[36(301t^2 - 119t + 12)\gamma^3 + 36(25722t^2 - 10393t + 1089)\gamma^2 + (14412383t^2 - 5957602t + 670296)\gamma + (275789t^2 + 6336034t - 709932)]/1953125.$$

Observe that all coefficients of γ'^i in the function $b(\gamma')$ are positive, $0 \leq i \leq 3$. Thus, $b(\gamma')$ is positive for all $\gamma' \in (0,1)$. It follows that $\beta(\gamma, x)$ is positive for all $\gamma \in (\gamma_{\beta(t)}, 1)$.

Just as $\alpha(\gamma)$ needs to be less or equal to 1 for all $\gamma \in (0, \gamma_{\beta(t)})$, proportion $\beta(\gamma)$ must be less or equal to 1 for all $\gamma \in (\gamma_{\beta(t)}, 1)$. The function $\beta(\gamma, x)$ does not exceed 1 if its numerator is less or equal to its denominator, i.e.,

$$5[(1-5t)x + t](1-\gamma^2)(\gamma + 3)$$
$$\leq 2[(11t-1)\gamma^3 + 30t\gamma^2 + (10t+1)\gamma - 3t]x + t(1-\gamma)(2\gamma^2 + 23\gamma + 27)$$
$$\Leftrightarrow \underbrace{[(t-1)\gamma^3 + 5(t-1)\gamma^2 - (15t-1)\gamma - 23t + 5]}_{<0}x - t(1-\gamma)(4-y-y^2) < 0$$

As $\beta(\gamma) = \beta(\gamma, x = x_{28})$ and $x_{28} > 0$ for all $\gamma \in (0.3, 1)$, proportion $\beta(\gamma) \leq 1$ for all $\gamma \in (\gamma_{\beta(t)}, 1)$. Hence, we arrive at sign $h'_1(x_{28}) \neq$ sign $h'_{28}(x_{28})$. ✓

Properties 1. through 8. prove Theorem 4 completely. □

4 Optimal Designs

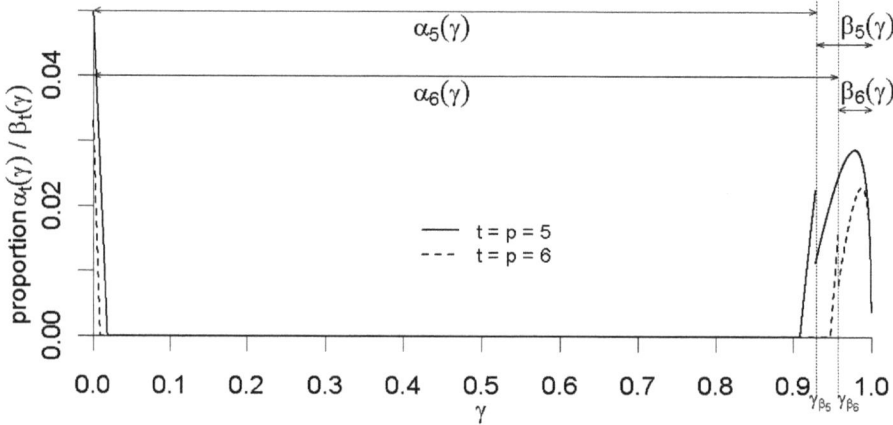

Figure 4.4: Sequence proportions $\alpha(\gamma)$ of equivalence class 2 and $\beta(\gamma)$ of equivalence classes 19 and 28, respectively, for an approximate optimal design with $t = p = 5$ and $t = p = 6$ periods.

5 Conclusions and Recommendations

The structure of optimal crossover designs is the main topic of this thesis. We investigate a model with carryover effects and, additionally, with interaction effects between products and assessors. Although there are many papers dealing with the construction of optimal designs including carryover effects, the number of authors dealing with models which include additional interaction effects is limited, e.g. Kemmler (1990).

A method by Kushner (1997) serves as a tool, in which the trace of the information matrix of the design is maximized on the basis of design-dependent equivalence class functions h_l. The equivalence class functions h_l are determined by the sequence of treatments which is given to an assessor. Since 3-, 4-, 5-, and 6- periods experiments are the most practical sequence lengths used in sensory studies, this thesis covers those four cases for treatment sequences. As a result, it is possible to extract three sequence classes l of which either two of their h_l functions or all three of them present the minimum of the maximal equivalence class function. The three sequences are determined by $[1, 2, \ldots, p-1, p]$, $[1, 2, \ldots, p-1, p-1]$ and $[1, 2, 2, 3, 3, \ldots, \frac{p+1}{2}, \frac{p+1}{2}]$ for odd p, or $[1, 2, 3, 3, \ldots, \frac{p+2}{2}, \frac{p+2}{2}]$ for even p, respectively.

Dependent on the coefficient of the variance of the interaction $\gamma \in (0, 1)$, the proportion of the three sequences in an approximate optimal design can vary tremendously:

Example 1: If $t = 4$, $p = 4$ and $\gamma = 0.2$, the optimal design only consists of sequences of class $[1, 2, 3, 4]$ and could be the following Latin square: $\begin{bmatrix} 1 & 2 & 3 & 4 \\ 2 & 4 & 1 & 3 \\ 3 & 1 & 4 & 2 \\ 4 & 3 & 2 & 1 \end{bmatrix}$. This optimal design fulfills the conditions of section 2.2.2, i.e., $C_f = C_f^{(M1)}$ and its partitioned matrices C_{fij}, $1 \leq i, j \leq 2$, are completely symmetric.

Example 2: If $t = 4$, $p = 4$ and $\gamma = 0.9$, about 6.5% of the sequences of an optimal design are representatives of class $[1, 2, 3, 3]$. All other sequences of the optimal design are to be chosen

5 Conclusions and Recommendations

from class $[1, 2, 3, 4]$.
Unfortunately, the proportion of sequences of class $[1, 2, 3, 3]$ can only be realized at 0% or 25%. Comparing the traces of the information matrix of the two possible designs, the first has an efficiency of 99.35%, whereas the second has an efficiency of about 94.01% if 25% of its sequences are selected from class $[1, 2, 3, 3]$.

Example 3: If $t = p = 5$ and $\gamma = 0.95$, about 2.2% of the sequences of an optimal design are representatives of class $[1, 2, 2, 3, 3]$, all other sequences are representatives of class $[1, 2, 3, 4, 5]$.

It is not unusual that proportions of about 2.2% of sequence classes are not realizable, considering the rather small number of assessors in an experiment. Referring to Example 3, there have to be $n = 46$ assessors in order to keep the specified proportions of sequences, in which one sequence is a representative of $[1, 2, 2, 3, 3]$ and 45 sequences are representatives of class $[1, 2, 3, 4, 5]$. Efficiency calculations show, that designs, which fulfill the advised proportions of sequences, have a higher efficiency as designs which do not repeat treatments. Therefore, small, not realizable proportions can be interpreted as an indicator for including at least one representative of the denoted sequence class to get a better, unbiased estimator of the treatments and their carryover effects in the specified model. However, as Example 2 shows, if the recommended proportion of sequence is highly exceeded, a design with no repetition of treatments performs a lot better.

As a general conclusion for sequence lengths $3 \leq p \leq 6$, an approximate optimal design is a combination of sequences of three equivalence classes: $1 : [1, 2, \ldots, p-1, p]$, $2 : [1, 2, \ldots, p-1, p-1]$, and $3a : [1, 2, 2, 3, 3, \ldots, \frac{p+1}{2}, \frac{p+1}{2}]$ for odd p, or $3b : [1, 2, 3, 3, \ldots, \frac{p+2}{2}, \frac{p+2}{2}]$ for even p, respectively. The proportion of sequences of each equivalence class of an approximate optimal design is dependent on the parameter γ. If $p = 3$ or $p = 4$ equivalence classes 2 and $3a$, or 2 and $3b$, are identical since their specifications of the treatment arrangements are the same. In contrast to the prediction of the introduction in chapter 1, the proportion of sequences of equivalence class 1 does not increase on the entire interval $(0, 1)$ as the coefficient γ increases. At some point of γ getting close to 1, the partitioned matrix C_{d22} increases extremely, such that the stated proportion decreases again and sequences of equivalence class 2 and k get more important. An explanation could be that the difference of the two replicated observations is an estimator of the corresponding carryover effect, with an extremely small variance if γ is close to 1. Still, the proportion of sequences of equivalence class 2 does not exceed its magnitude in the traditional model with carryover effects and no unit×treatment interaction. To summarize, this thesis illustrates that it is important to model interaction and carryover effects separately if unit×treatment interaction is likely. The results for the structure of optimal crossover designs of the assumed model with interaction and carryover effects demonstrate their difference to the optimal designs of the traditional model without unit×treatment interaction, even though the efficiency of optimal designs of the traditional model is very high.

The global intention of this thesis could be the perspective of a generalization of the optimal findings for a general number of periods p. The challenge is, however, the proper treatment of the "nuisance" parameter γ.

5 Conclusions and Recommendations

Appendix A : Notations

General Variables:

γ	$\in (0,1)$, coefficient of the variance of $w_{d(u,r),u}$ for identical treatments $d(u,r)$ given to the same subjects u
$u = 1,\ldots,n$	units/subjects of the experiment
$r = 1,\ldots,p$	periods of the experiment
$i = 1,\ldots,t$	treatments of the experiment
$i = 1,\ldots,t^*$	treatments in one unit
$l = 1,\ldots,K$	equivalence classes of sequences
u_l	sequence of treatments of class l
$n_j(l)$	is the number of appearances of treatment j in the sequence
$\tilde{n}_j(l)$	is the number of appearances of the carryover effect j in the sequence (including itself)
$\tilde{n}_{ij}(l)$	number of appearances of treatment j following treatment i in the sequence, whereas $\tilde{n}_{jj}(l)$ is the number of appearances of treatment j following itself
$\tilde{n}_{0j}(l)$	$= 1$, if treatment j is in the first period; 0 otherwise
h_l	equivalence class function of sequence class l
x_{min}	location of the minimum of the $h_1(x)$ parabola
$\gamma_{\beta(t)}$	parameter function of t in which $h_2(x_2) = h_k(x_k)$ and $x_2 = x_k$

Appendix A : Notations

Math Analogy:

$\otimes$	Kronecker product
I_n	n-dimensional identity matrix
1_n	n-vector of ones
A^-	generalized inverse (g-inverse) of matrix A
$[A, B]$	matrix with partitions A and B
$\operatorname{tr} A$	trace of matrix A: sum of A's diagonal elements
$\omega(A) = A(A^T A)^- A^T$	projection onto column space of matrix A
$\omega^\perp(A) = I - \omega(A)$	projection onto the orthogonal column space of matrix A
$B_t = I_t - \frac{1}{t} 1_t 1_t^T$	centralizing matrix
$a_{n_j(l)} = \frac{(n_j(l)-2)\gamma+1}{[(n_j(l)-1)\gamma+1](1-\gamma)}$	diagonal entry of matrix S_{du}^{-1}, $\forall n_j(l) \geq 1$
$b_1 := 0$	off-diagonal entry of matrix S_{du}^{-1} for $n_j(l) = 1$
$b_{n_j(l)} = -\frac{\gamma}{[(n_j(l)-1)\gamma+1](1-\gamma)}$	off-diagonal entry of matrix S_{du}^{-1}, $\forall n_j(l) > 1$
$crs_{n_j(l)} = a_{n_j(l)} + (n_j(l) - 1)b_{n_j(l)}$	column- or row- sum of S_{du}^{-1}

Design-Dependent Objects:

$\Omega_{t,n,p}$	the set of all designs with t treatments, n subjects and $p \leq t$ periods
$d(u,r)$	treatment of design d assigned to period r of unit u
$S_d = diag(S_{d1}, \ldots, S_{dn})$	diagonal matrix with elements S_{du}
$S_{du} = I_p + \Sigma_{du}$	covariance factor of $e_{u,r} + w_{d(u,r),u}$
Σ_{du}	covariance matrix of the interaction effects of unit u
$e_{u,r}$	(random) error, $1 \leq u \leq n, 1 \leq r \leq p$, in model (M0)
$w_{d(u,r),u}$	(random) interaction effect between treatment $d(u,r)$ and unit u
$V_d = diag(V_{d1}, \ldots, V_{dn})$	diagonal matrix with elements V_{du} and properties: $V_d V_d = S_d^{-1}$ and $V_d S_d V_d = I_{np}$
$V_d^* = V_d \omega^\perp (V_d 1_p) V_d$	diagonal matrix with elements V_{du}^*, cf. equation (2.3)
$R_u = \sum_{r=1}^{p} crs_{n_j(l),r}$ $= \sum_{i=1}^{t} n_j(l) crs_{n_j(l)}$	sum of all entries of S_{du}^{-1}

Appendix A : Notations

Appendix B : Equivalence Classes of Treatments

B.1 Sequence Length $p = 5$

B.1.1 Set of all Equivalence Classes

$l : [...]$	$h_l =$	$c_{11}(l)$	$+2c_{12}(l)x$	$+c_{22}(l)x^2$
$R_1 = 5$				
1 : [12345]	$h_1 =$	4	$-\frac{8}{5} \cdot x$	$+\frac{4(4t-1)}{5t} \cdot x^2$
Group A: $R_A = (3\gamma+5)/(\gamma+1)$				
2 : [12344]	$h_2 =$	$\frac{6(\gamma+3)}{(3\gamma+5)}$	$-\frac{4\gamma}{(3\gamma+5)} \cdot x$	$+\frac{(-(4t-2)\gamma^2-(4t-2)\gamma+16t-4)}{t(1-\gamma)(3\gamma+5)} \cdot x^2$
6 : [12334]	$h_6 =$	$\frac{6(\gamma+3)}{(3\gamma+5)}$	$-\frac{2(\gamma+1)}{(3\gamma+5)} \cdot x$	$+\frac{(-(4t-2)\gamma^2-(2t-2)\gamma+14t-4)}{t(1-\gamma)(3\gamma+5)} \cdot x^2$
18 : [12234]				
38 : [11234]	$h_{38} =$	$\frac{6(\gamma+3)}{(3\gamma+5)}$	$-\frac{2(2\gamma+1)}{(3\gamma+5)} \cdot x$	$+\frac{(-6t\gamma^2-4t\gamma+14t-4)}{t(1-\gamma)(3\gamma+5)} \cdot x^2$
3 : [12343]	$h_3 =$	$\frac{6(\gamma+3)}{(3\gamma+5)}$	$-\frac{2(\gamma+5)}{(3\gamma+5)} \cdot x$	$+\frac{(-(4t-2)\gamma^2-(4t-2)\gamma+16t-4)}{t(1-\gamma)(3\gamma+5)} \cdot x^2$
4 : [12342]				
5 : [12341]	$h_5 =$	$\frac{6(\gamma+3)}{(3\gamma+5)}$	$-\frac{2(2\gamma+5)}{(3\gamma+5)} \cdot x$	$+\frac{(-6t\gamma^2-6t\gamma+16t-4)}{t(1-\gamma)(3\gamma+5)} \cdot x^2$
10 : [12324]	$h_{10} =$	$\frac{6(\gamma+3)}{(3\gamma+5)}$	$-\frac{12}{(3\gamma+5)} \cdot x$	$+\frac{(-(2t-2)\gamma^2-(4t-2)\gamma+14t-4)}{t(1-\gamma)(3\gamma+5)} \cdot x^2$
14 : [12314]	$h_{14} =$	$\frac{6(\gamma+3)}{(3\gamma+5)}$	$-\frac{2(\gamma+6)}{(3\gamma+5)} \cdot x$	$+\frac{(-4t\gamma^2-6t\gamma+14t-4)}{t(1-\gamma)(3\gamma+5)} \cdot x^2$
28 : [12134]				

Table B.1 continues on the next page ...

Appendix B : Equivalence Classes of Treatments

... continued from the previous page.

$l : [\ldots]$	$h_l =$	$c_{11}(l)$	$+2c_{12}(l)x$	$+c_{22}(l)x^2$
Group B: $R_B = (\gamma+5)/(\gamma+1)$				
$k \equiv 19 : [12233]$	$h_{19} =$	$\frac{8(\gamma+2)}{(\gamma+1)(\gamma+5)}$	$+\frac{2(3+\gamma)}{(\gamma+1)(\gamma+5)} \cdot x$	$+\frac{(4\gamma^2+10t\gamma+14t-4)}{t(1-\gamma)(\gamma+1)(\gamma+5)} \cdot x^2$
$39 : [11233]$	$h_{39} =$	$\frac{8(\gamma+2)}{(\gamma+1)(\gamma+5)}$	$+\frac{2(3-\gamma)}{(\gamma+1)(\gamma+5)} \cdot x$	$+\frac{(-2t\gamma^2+(6t-2)\gamma+14t-4)}{t(1-\gamma)(\gamma+1)(\gamma+5)} \cdot x^2$
$42 : [11223]$	$h_{42} =$	$\frac{8(\gamma+2)}{(\gamma+1)(\gamma+5)}$	$+\frac{4}{(\gamma+1)(\gamma+5)} \cdot x$	$+\frac{(-2t\gamma^2+(8t-2)\gamma+12t-4)}{t(1-\gamma)(\gamma+1)(\gamma+5)} \cdot x^2$
$8 : [12332]$	$h_8 =$	$\frac{8(\gamma+2)}{(\gamma+1)(\gamma+5)}$	$-\frac{4}{(\gamma+1)(\gamma+5)} \cdot x$	$+\frac{(4\gamma^2+10t\gamma+14t-4)}{t(1-\gamma)(\gamma+1)(\gamma+5)} \cdot x^2$
$9 : [12331]$ $40 : [11232]$	$h_9 =$	$\frac{8(\gamma+2)}{(\gamma+1)(\gamma+5)}$	$-\frac{4}{(\gamma+5)} \cdot x$	$+\frac{(-4t\gamma^2+(8t-2)\gamma+14t-4)}{t(1-\gamma)(\gamma+1)(\gamma+5)} \cdot x^2$
$21 : [12231]$ $29 : [12133]$	$h_{21} =$	$\frac{8(\gamma+2)}{(\gamma+1)(\gamma+5)}$	$-\frac{4}{(\gamma+5)} \cdot x$	$+\frac{(-2t\gamma^2+(6t-2)\gamma+14t-4)}{t(1-\gamma)(\gamma+1)(\gamma+5)} \cdot x^2$
$25 : [12213]$	$h_{25} =$	$\frac{8(\gamma+2)}{(\gamma+1)(\gamma+5)}$	$-\frac{2(\gamma+3)}{(\gamma+1)(\gamma+5)} \cdot x$	$+\frac{(-2t\gamma^2+(8t-2)\gamma+12t-4)}{t(1-\gamma)(\gamma+1)(\gamma+5)} \cdot x^2$
$11 : [12323]$	$h_{11} =$	$\frac{8(\gamma+2)}{(\gamma+1)(\gamma+5)}$	$-\frac{2(\gamma+7)}{(\gamma+1)(\gamma+5)} \cdot x$	$+\frac{(-(2t-4)\gamma^2+14t-4)}{t(1-\gamma)(\gamma+1)(\gamma+5)} \cdot x^2$
$13 : [12321]$ $30 : [12132]$	$h_{13} =$	$\frac{8(\gamma+2)}{(\gamma+1)(\gamma+5)}$	$-\frac{2(3\gamma+7)}{(\gamma+1)(\gamma+5)} \cdot x$	$+\frac{(-2t\gamma^2+(6t-2)\gamma+14t-4)}{t(1-\gamma)(\gamma+1)(\gamma+5)} \cdot x^2$
$16 : [12312]$	$h_{16} =$	$\frac{8(\gamma+2)}{(\gamma+1)(\gamma+5)}$	$-\frac{2(3\gamma+7)}{(\gamma+1)(\gamma+5)} \cdot x$	$+\frac{(-6t\gamma^2-(2t+2)\gamma+14t-4)}{t(1-\gamma)(\gamma+1)(\gamma+5)} \cdot x^2$
$15 : [12313]$	$h_{15} =$	$\frac{8(\gamma+2)}{(\gamma+1)(\gamma+5)}$	$-\frac{2(5\gamma+7)}{(\gamma+1)(\gamma+5)} \cdot x$	$+\frac{(-2t\gamma^2+(6t-2)\gamma+14t-4)}{t(1-\gamma)(\gamma+1)(\gamma+5)} \cdot x^2$
$32 : [12123]$	$h_{32} =$	$\frac{8(\gamma+2)}{(\gamma+1)(\gamma+5)}$	$-\frac{4(\gamma+4)}{(\gamma+1)(\gamma+5)} \cdot x$	$+\frac{(-4t\gamma^2-(2t+2)\gamma+12t-4)}{t(1-\gamma)(\gamma+1)(\gamma+5)} \cdot x^2$

Table B.1 continues on the next page ...

B.1 Sequence Length $p = 5$

... continued from the previous page.

$l : [...]$	$h_l =$	$c_{11}(l)$	$+2c_{12}(l)x$	$+c_{22}(l)x^2$
Group C: $R_C = (4\gamma+5)/(2\gamma+1)$				
7 : [12333]	$h_7 =$	$\frac{2(2\gamma+7)}{(4\gamma+5)}$	$+\frac{4(1-\gamma)}{(4\gamma+5)} \cdot x$	$+\frac{(-(2t-2)\gamma^2+2\gamma+14t-4)}{t(1-\gamma)(4\gamma+5)} \cdot x^2$
22 : [12223]	$h_{22} =$	$\frac{2(2\gamma+7)}{(4\gamma+5)}$	$+0$	$+\frac{(-(2t-2)\gamma^2+(4t+2)\gamma+10t-4)}{t(1-\gamma)(4\gamma+5)} \cdot x^2$
48 : [11123]	$h_{48} =$	$\frac{2(2\gamma+7)}{(4\gamma+5)}$	$-\frac{4\gamma}{(4\gamma+5)} \cdot x$	$+\frac{(-4t\gamma^2-2\gamma+10t-4)}{t(1-\gamma)(4\gamma+5)} \cdot x^2$
12 : [12322] 20 : [12232]	$h_{12} =$	$\frac{2(2\gamma+7)}{(4\gamma+5)}$	$-\frac{6}{(4\gamma+5)} \cdot x$	$+\frac{(-(2t-2)\gamma^2+(6t+2)\gamma+14t-4)}{t(1-\gamma)(4\gamma+5)} \cdot x^2$
17 : [12311] 41 : [11231]	$h_{17} =$	$\frac{2(2\gamma+7)}{(4\gamma+5)}$	$-\frac{2(2\gamma+3)}{(4\gamma+5)} \cdot x$	$+\frac{(-4t\gamma^2+(2t-2)\gamma+14t-4)}{t(1-\gamma)(4\gamma+5)} \cdot x^2$
35 : [12113] 45 : [11213]	$h_{35} =$	$\frac{2(2\gamma+7)}{(4\gamma+5)}$	$-\frac{10}{(4\gamma+5)} \cdot x$	$+\frac{((2t-2)\gamma+10t-4)}{t(1-\gamma)(4\gamma+5)} \cdot x^2$
31 : [12131]	$h_{31} =$	$\frac{2(2\gamma+7)}{(4\gamma+5)}$	$-\frac{16}{(4\gamma+5)} \cdot x$	$+\frac{(-(2t+2)\gamma+14t-4)}{t(1-\gamma)(4\gamma+5)} \cdot x^2$
Group D: $R_D = (7\gamma+5)/((\gamma+1)(2\gamma+1))$				
43 : [11222]	$h_{43} =$	$\frac{12}{(7\gamma+5)}$	$+\frac{10}{(7\gamma+5)} \cdot x$	$+\frac{((10t-2)\gamma+12t-4)}{t(1-\gamma)(7\gamma+5)} \cdot x^2$
49 : [11122]	$h_{49} =$	$\frac{12}{(7\gamma+5)}$	$+\frac{8}{(7\gamma+5)} \cdot x$	$+\frac{((10t-4)\gamma+10t-4)}{t(1-\gamma)(7\gamma+5)} \cdot x^2$
27 : [12211] 44 : [11221]	$h_{27} =$	$\frac{12}{(7\gamma+5)}$	$+0$	$+\frac{((16t-4)\gamma+12t-4)}{t(1-\gamma)(7\gamma+5)} \cdot x^2$
24 : [12221]	$h_{24} =$	$\frac{12}{(7\gamma+5)}$	$-\frac{2}{(7\gamma+5)} \cdot x$	$+\frac{((12t-2)\gamma+10t-4)}{t(1-\gamma)(7\gamma+5)} \cdot x^2$
26 : [12212] 33 : [12122]	$h_{26} =$	$\frac{12}{(7\gamma+5)}$	$-\frac{10}{(7\gamma+5)} \cdot x$	$+\frac{((10t-2)\gamma+12t-4)}{t(1-\gamma)(7\gamma+5)} \cdot x^2$
36 : [12112]	$h_{36} =$	$\frac{12}{(7\gamma+5)}$	$-\frac{12}{(7\gamma+5)} \cdot x$	$+\frac{((6t-4)\gamma+10t-4)}{t(1-\gamma)(7\gamma+5)} \cdot x^2$

Table B.1 continues on the next page ...

Appendix B : Equivalence Classes of Treatments

... continued from the previous page.

$l : [...]$	$h_l =$	$c_{11}(l)$	$+2c_{12}(l)x$	$+c_{22}(l)x^2$
46 : [11212]				
34 : [12121]	$h_{34} =$	$\frac{12}{(7\gamma+5)}$	$-\frac{20}{(7\gamma+5)} \cdot x$	$+\frac{(-(4t+4)\gamma+12t-4)}{t(1-\gamma)(7\gamma+5)} \cdot x^2$
Group E: $R_E = (3\gamma+5)/(3\gamma+1)$				
23 : [12222]	$h_{23} =$	$\frac{8}{(3\gamma+5)}$	$+\frac{4}{(3\gamma+5)} \cdot x$	$+\frac{((2t+4)\gamma+10t-4)}{t(1-\gamma)(3\gamma+5)} \cdot x^2$
51 : [11112]	$h_{51} =$	$\frac{12}{(3\gamma+5)}$	$-\frac{2}{(3\gamma+5)} \cdot x$	$+\frac{((2t-2)\gamma+4t-4)}{t(1-\gamma)(3\gamma+5)} \cdot x^2$
37 : [12111]	$h_{37} =$	$\frac{12}{(3\gamma+5)}$	$-\frac{6}{(3\gamma+5)} \cdot x$	$+\frac{((4t-2)\gamma+10t-4)}{t(1-\gamma)(3\gamma+5)} \cdot x^2$
47 : [11211]				
50 : [11121]				
$R_{52} = 5/(4\gamma+1)$				
52 : [11111]	$h_{52} =$	0	$+0$	$+\frac{4(t-1)}{5t(1-\gamma)} \cdot x^2$

Table B.1: All equivalence classes, their representative sequences [...] and h_l functions for sequence length $p = 5$.

B.1.2 Steps of Argumentation

As Proof of Lemma 7. ...

b) Next to analyze is

$$(h_{19} - h_7)(x) = \frac{2(1-\gamma)(2\gamma^2 + 5\gamma + 5)}{(\gamma+1)(\gamma+5)(4\gamma+5)} + \frac{2(2\gamma^3 + 14\gamma^2 + 15\gamma + 5)}{(\gamma+1)(\gamma+5)(4\gamma+5)} x$$
$$\frac{2\gamma[(t-1)\gamma^3 + (6t+1)\gamma^2 + (18t+1)\gamma + 11t - 1]}{t(1-\gamma)(\gamma+1)(\gamma+5)(4\gamma+5)} x^2.$$

All ratios of $(h_{19} - h_7)$ are positive because its factors are positive for $\gamma \in (0, 1)$ and all $t \geq 5$. It follows that $h_{19} > h_7$ for all $x > 0$.

c) Analyze

$$(h_{19} - h_{12})(x) = \frac{2(1-\gamma)(2\gamma^2 + 5\gamma + 5)}{(\gamma+1)(\gamma+5)(4\gamma+5)} + \frac{2(7\gamma^2 + 35\gamma + 30)}{(\gamma+1)(\gamma+5)(4\gamma+5)}x$$
$$+ \frac{2\gamma[(1-t)\gamma^2 - 4t\gamma - 4t - 1]}{t(\gamma+1)(\gamma+5)(4\gamma+5)}x^2$$

and write $(h_{19} - h_{12})(x) = dc_{11} + dc_{12}x + dc_{22}x^2$ to observe that $dc_{11} > 0$, $dc_{12} > 0$ and $dc_{22} < 0$. Since $x^2 < x$ for all $x \in (0,1)$, we get $dc_{12}x + dc_{22}c^2 > (dc_{12} + dc_{22})x^2 > 0$ for all $\gamma \in (0,1)$ and all $t \geq 5$. It follows that $h_{19} > h_{12}$ for all $x \in (0,1)$.

Notice, the conclusion of b) and c) is: $h_{19}(x) > h_l(x)$ for $l \in L_C$, $\gamma \in (0,1)$ and $x \in (0,1)$.

d) Furthermore,

$$(h_{19} - h_{43})(x) = \frac{4(11\gamma^2 + 20\gamma + 5)}{(\gamma+1)(\gamma+5)(7\gamma+5)} - \frac{4(5 + 2\gamma - \gamma^2)}{(\gamma+1)(\gamma+5)(7\gamma+5)}x$$
$$+ \frac{2[(15 - 5t)\gamma^3 + (18 - t)\gamma^2 + (13t + 3)\gamma + 5t]}{t(\gamma-1)(\gamma+1)(\gamma+5)(7\gamma+5)}x^2.$$

Retype $h_{19} - h_{43}$ as $dc_{11} + dc_{12}x + dc_{22}x^2$. The coefficients dc_{11} and dc_{22} are positive, dc_{12} is negative and for $\gamma \in (0,1)$ and $t \geq 5$. Assume $x \in (0,1)$, and we get $dc_{11} + dc_{12}x > (dc_{11} + dc_{12})x > 0$. It follows that $h_{19}(x) > h_{43}(x)$ for all $x \in (0,1)$.

e) Next to examine is

$$(h_{19} - h_{27})(x) = \frac{4(11\gamma^2 + 20\gamma + 5)}{(\gamma+1)(\gamma+5)(7\gamma+5)} + \frac{2(3+\gamma)}{(\gamma+1)(\gamma+5)}x$$
$$+ \frac{2[(16 - 8t)\gamma^3 + (24 - 19t)\gamma^2 + (8 - 2t)\gamma + 5t]}{t(1-\gamma)(\gamma+1)(\gamma+5)(7\gamma+5)}x^2.$$

Substitution of $x = x_{19}$ provides

$$(h_{19} - h_{27})(x_{19}) = \frac{a(\gamma) + b(\gamma)\sqrt{RT_{x_{19}}}}{(7\gamma+5)(8t\gamma^3 - 2\gamma^3 + 40t\gamma^2 + 17t\gamma + 2\gamma - 5t)^2},$$

in which

$a(\gamma) = 8(236t^2 + 145t - 58)\gamma^6 + 2(6388t^2 + 3981t - 208)\gamma^5 + 4(5093t^2 + 2145t + 96)\gamma^4 + 2(2697t^2 + 2524t + 208)\gamma^3 - 10(1033t^2 - 450t - 8)\gamma^2 - 50t(61t - 31)\gamma + 1750t^2$

and

$b(\gamma) = 2(44t - 39)\gamma^3 + 4(67t - 36)\gamma^2 + 10(7t - 5)\gamma - 50t$.

The denominator of $(h_{19} - h_{27})(x_{19})$ is positive for all $\gamma \in (0,1)$ and $t \geq 5$. In order to determine the roots of the numerator, the signs of $a(\gamma)$ and $b(\gamma)$ need to be identified.

81

Appendix B : Equivalence Classes of Treatments

e1) In order to analyze $a(\gamma)$ for all $\gamma > \gamma_{\beta(t)}$, transform γ by $\gamma = 0.08\gamma' + 0.92$. It follows that $a(\gamma) > 0$ is equivalent to $a(\gamma') > 0$ for $\gamma' \in (0,1)$ and $\gamma \in (0.92, 1)$. We get

$a(\gamma') = \frac{16}{244140625}[32(236t^2 + 145t - 58)\gamma'^6 + 4(289972t^2 + 179565t - 37216)\gamma'^5 + 10(6444478t^2 + 3745785t - 463784)\gamma'^4 + 5(340472317t^2 + 190747390t - 13934576)\gamma'^3 + 10(2224492503t^2 + 1332710235t - 49288784)\gamma'^2 + (132229285063t^2 + 100682696410t - 1129648864)\gamma' + 2(141605470551t^2 + 162225704270t + 848499072)]$.

As all coefficients of γ'^i are positive, $0 \leq i \leq 6$, $a(\gamma')$ is positive for all $\gamma' \in (0,1)$, and thus, $a(\gamma)$ is positive for all $\gamma \in (0.92, 1)$ as well.

e2) Abbreviate $b(\gamma)$ as $b_3\gamma^3 + b_2\gamma^2 + b_1\gamma + b_0$. The first derivative of $b(\gamma)$ is $b'(\gamma) = 3b_3\gamma^2 + 2b_2\gamma + b_1 > 0$ for all $\gamma \in (0,1)$ and $t \geq 5$, since b_1, b_2 and b_3 are positive. A positive slope denotes that $b(\gamma)$ increases monotonously in γ. Therefore, as $b(0.92) > 1320$ is positive, $b(\gamma)$ is positive for all $\gamma \in (0.92, 1)$.

Use the results of e1) and e2) to find that $a(\gamma) + b(\gamma)\sqrt{RT_{x_{19}}}$ is positive for all $\gamma \in (0.92, 1)$. The inequality implies that $h_{19}(x_{19}) > h_{27}(x_{19})$ for all $\gamma \in [\gamma_{\beta(t)}, 1)$.
Notice, combining the results of d) and e), we get $h_{19}(x_{19}) > h_l(x_{19})$ for all $l \in L_D$ and $\gamma \in [\gamma_{\beta(t)}, 1)$.

f) Now, observe that

$$(h_{19} - h_{23})(x) = \frac{8(2\gamma^2 + 5\gamma + 5)}{(\gamma + 1)(\gamma + 5)(3\gamma + 5)} + \frac{2(\gamma^2 + 2\gamma + 5)}{(\gamma + 1)(\gamma + 5)(3\gamma + 5)}x$$
$$\frac{2[(4-t)\gamma^3 + 4t\gamma^2 + (11t - 4)\gamma + 10t]}{t(1-\gamma)(\gamma + 1)(\gamma + 5)(3\gamma + 5)}x^2.$$

Write $(h_{19} - h_{23})(x) = dc_{11} + dc_{12}x + dc_{22}x^2$. It is clear to detect that dc_{11}, dc_{12} and dc_{22} are positive for $\gamma \in (0,1)$ and $t \geq 5$.
The positive coefficients of x and x^2 imply that $h_{19}(x) > h_{23}(x)$ for all $x > 0$.

g) Next to analyze is

$$(h_{19} - h_{37})(x) = \frac{8(2\gamma^2 + 5\gamma + 5)}{(\gamma + 1)(\gamma + 5)(3\gamma + 5)} + \frac{4(3\gamma^2 + 16\gamma + 15)}{(\gamma + 1)(\gamma + 5)(3\gamma + 5)}x$$
$$\frac{2[(7-2t)\gamma^3 + (18 - 2t)\gamma^2 + (6t + 11)\gamma + 10t]}{t(1-\gamma)(\gamma + 1)(\gamma + 5)(3\gamma + 5)}x^2.$$

B.1 Sequence Length $p = 5$

Again, as in f), all the coefficients dc_{ij} are positive, $1 \leq i, j \leq 2$, for $\gamma \in (0,1)$ and $t \geq 5$. The conclusion is: $h_{19}(x) > h_{37}(x)$ for all $x > 0$.

Notice, summing up f) and g), it follows that $h_{19} > h_l$ for all $l \in L_E \cup \{52\}$, $\gamma \in (0,1)$ and $x > 0$.

$\square$

As Proof of Lemma 8. ...

c) Remaining

$$(h_2 - h_{27})(x) = \frac{6(7\gamma^2 + 20\gamma + 5)}{(3\gamma + 5)(7\gamma + 5)} - \frac{4\gamma}{(3\gamma + 5)}x$$
$$+ \frac{2[7(1 - 2t)\gamma^3 + 6(3 - 8t)\gamma^2 + (7 - 12t)\gamma + 10t]}{t(1 - \gamma)(3\gamma + 5)(7\gamma + 5)}x^2$$

to be analyzed properly. Substitution of $x = x_2$ supplies

$$(h_2 - h_{27})(x_2) = \frac{a(\gamma) + b(\gamma)\sqrt{RT_{x_2}}}{\gamma^2(7\gamma + 5)(14t\gamma - \gamma + 6t + 1)^2},$$

in which
$a(\gamma) = \ 4[14(12t^2 + 13t - 1)\gamma^5 - (12t^2 - 356t + 7)\gamma^4 + (1208t^2 - 97t + 16)\gamma^3 -$
$(274t^2 - 214t - 5)\gamma^2 - 5t(98t - 29)\gamma + 200t^2]$

and

$b(\gamma) = \ 8[-14t\gamma^3 + 3(12t - 5)\gamma^2 + (17t - 7)\gamma - 10t].$

The denominator of $(h_2 - h_{27})(x_2)$ is positive for all $\gamma \in (0,1)$ and $t \geq 5$. The next step is the determination of the roots of $a(\gamma) + b(\gamma)\sqrt{RT_{x_2}}$ in the interval $(0, 0.93) \supset (0, \gamma_{\beta(t)}]$ of γ.

c1) Abbreviate $a(\gamma) = a_5\gamma^5 + a_4\gamma^4 + a_3\gamma^3 + a_2\gamma^2 + a_1\gamma + a_0$ and decompose a_3 into $48t^2 + (4784t^2 - 388t + 64)$. Next, define $A_{03}(\gamma) := a_0 + a_1\gamma + a_2\gamma^2 + (4784t^2 - 388t + 64)\gamma^3$. The first derivative of $A_{03}(\gamma)$ is $A'_{03}(\gamma) = 12(1196t^2 - 97t + 16)\gamma^2 + 2a_2\gamma + a_1$ and equals 0 iff $\gamma \in (0,1)$ is equivalent to

$$\gamma_{st} = \frac{\sqrt{W} + 274t^2 - 214t - 5}{3(1196t^2 - 97t + 16)},$$

Appendix B : Equivalence Classes of Treatments

in which $W = 1833196t^4 - 780122t^3 + 108771t^2 - 4820t + 25$. Some simple calculus confirms that $0.4 < \gamma_{st} < 0.5$. Analyzing the curvature of $A_{03}(\gamma)$, the root of $A_{03}''(\gamma)$ is located at some $\gamma_{inf} < 0.1$ such that $A_{03}((0, \gamma_{inf}))$ is concave and $A_{03}((\gamma_{inf}, 0.93))$ convex. This, however, proves $A_{03}(\gamma_{st})$ is a minimum of $A_{03}(\gamma)$. Remodel the function value of $A_{03}(\gamma_{st})$ properly and it proves $A_{03}(\gamma_{st}) > 0$. Knowing that $A_{03}(\gamma \searrow 0)$ is positive, the function $A_{03}(\gamma)$ is positive for all $\gamma \in (0, 0.93)$. In addition to $A_{03}(\gamma)$, $a_5 > 0$ and $a_4\gamma^4 + 48t^2\gamma^3 > (a_4 + 48t^2)\gamma^3 > 0$, such that $a(\gamma)$ is positive for all $\gamma \in (0, 0.93)$ and $t \geq 5$.

c2) In order to determine the range of γ in which $b(\gamma)$ is nonpositive, observe that $b'(\gamma) = -336t\gamma^2 + 48(12t - 5)\gamma + 8(17t - 7) > 0$ for $\gamma \in (0, 1)$ and $t \geq 5$ because $-336t\gamma^2 + 48(12t - 5)\gamma > 240(t - 1)\gamma^2 > 0$ for all $t \geq 5$. The positive slope of $b(\gamma)$ implies that there exists just one root in the interval $(0, 1)$ of γ, since $b(\gamma \searrow 0)$ is negative and $b(\gamma \nearrow 1)$ is positive. The function value $b(0.4)$ is positive. Thus, the true root of $b(\gamma)$ is at some $\gamma < 0.4$ and $b(\gamma)$ is positive for all $\gamma > 0.4$.

A positive function value of $b(\gamma)$ implies that $(h_2 - h_{27})(x_2)$ is positive. Thus, assume $b(\gamma) < 0$ and analyze if $a(\gamma) > -b(\gamma)\sqrt{RT_{x_2}}$. The inequality is equivalent to

$$a^2(\gamma) - b^2(\gamma)RT_{x_2} > 0$$
$$\Leftrightarrow 16[(14t - 1)\gamma + 6t + 1]^2 \cdot g(\gamma) > 0$$
$$\Leftrightarrow g(\gamma) := g_0 + g_1\gamma + \ldots + g_6\gamma^6 > 0,$$

in which

$g(\gamma) = 196(5t^2+2t+1)\gamma^6 - 28(194t^2-57t-21)\gamma^5 + (10680t^2-2636t+581)\gamma^4 + 2(516t^2+569t+105)\gamma^3 - 5(1363t^2-448t-5)\gamma^2 + 10t(53t+47)\gamma + 625t^2$.

It is sufficient to evaluate $g(\gamma)$ for $b(\gamma) < 0$, i.e., $\gamma \in (0, 0.4)$. Thus, restrict γ to interval $(0, 0.4)$ by the transformation $\gamma = 0.4\gamma'$, in which $\gamma' \in (0, 1)$. The transformation gives

$g(\gamma') = [12544(5t^2 + 2t + 1)\gamma'^6 - 4480(194t^2 - 57t - 21)\gamma'^5 + 400(10680t^2 - 2636t + 581)\gamma'^4 + 2000(516t^2 + 569t + 105)\gamma'^3 - 12500(1363t^2 - 448t - 5)\gamma'^2 + 62500t(53t + 47)\gamma' + 9765625t^2]/15625$.

Write $g(\gamma') = g_6'\gamma'^6 + \ldots + g_1'\gamma' + g_0'$. All g_i', $i \in \{0, 1, 3, 4, 6\}$, are positive, except for $i = 2, 5$. Define $G_{04}(\gamma') := \sum_{i=0}^{3} g_i'\gamma'^i + (g_4' + g_5')\gamma'^4$ to get $G_{04}(\gamma') < \sum_{i=0}^{5} g_i'\gamma'^i < g(\gamma')$. Take the derivative of $G_{04}(\gamma')$ twice and it yields $G_{04}'(\gamma') = 4(g_4'+g_5')\gamma'^3 + 3g_3'\gamma'^2 + 2g_2'\gamma' + g_1'$ and $G_{04}''(\gamma') = 12(g_4' + g_5')\gamma'^2 + 6g_3'\gamma'2g_2'$. The second derivative equals 0 iff γ' is equivalent

to
$$\gamma_{inf} = \frac{25\sqrt{3W} - 75(516t^2 + 569t + 105)}{12(42536t^2 - 9988t + 4081)},$$
in which $W = 116751904t^4 - 63577920t^3 + 20945057t^2 - 3198226t - 7735$. $G''_{04}(\gamma')$ is negative for all $\gamma' < \gamma_{inf}$. Thus, $G_{04}((0, \gamma_{inf}))$ is concave. The concavity implies that local minima of $G_{04}(\gamma)$ are represented in the range boundaries of either $\gamma' = 0$ or $\gamma' = \gamma_{inf}$. Since the function value of $G_{04}(\gamma' \searrow 0)$ is positive and some simple calculus confirms that $G_{04}(\gamma_{inf})$ is positive as well, it is concluded that $G_{04}(\gamma') > 0$ for all $\gamma' \in (0, \gamma_{inf})$. $G_{04}((\gamma_{inf}, 1))$ is convex. The convexity implies that $G'_{04}(\gamma)$ increases on the whole range of $\gamma' \in (\gamma_{inf}, 1)$. $G'_{04}(1)$ is negative, as well as G'_{04} for all $\gamma' \in (\gamma_{inf}, 1)$. This denotes $G_{04}(\gamma)$ is (monotonous) decreasing in this described interval of γ'. Observe that $G_{04}(1)$ is positive for all $t \geq 5$ and combine the result with the former declarations, and thus, $G_{04}(\gamma)$ is positive for all $\gamma' \in (\gamma_{inf}, 1)$.

To summarize, $G_{04}(\gamma)$ is positive for all $\gamma' \in (0, 1)$ and all $t \geq 5$. Hence, since $0 < G_{04}(\gamma') < g(\gamma')$, we get $(h_2 - h_{27})(x_2) > 0$.

Summing up the results of $l = 27$ and $l = 43$, $h_2(x_2) > h_l(x_2)$ for all $l \in L_D$, $\gamma \in (0, \gamma_{\beta(t)}]$ and $t \geq 5$.

□

Appendix B : Equivalence Classes of Treatments

B.2 Sequence Length $p = 6$

B.2.1 Set of all Equivalence Classes

$l : [\ldots]$	$h_l = c_{11}(l)$	$+2c_{12}(l)x$	$+c_{22}(l)x^2$
$R_1 = 6$			
$1 : [123456]$	5	$-\frac{5}{3}x$	$+\frac{5(5t-1)}{6t}x^2$
Group A: $R_A = 2(2\gamma+3)/(\gamma+1)$			
$2 : [123455]$	$\frac{2(3\gamma+7)}{(2\gamma+3)}$	$-\frac{3\gamma}{(2\gamma+3)}x$	$+\frac{((3-9t)\gamma^2+(2-6t)\gamma+25t-5)}{2t(1-\gamma)(2\gamma+3)}x^2$
$7 : [123445]$	$\frac{2(3\gamma+7)}{(2\gamma+3)}$	$-\frac{(2\gamma+1)}{(2\gamma+3)}x$	$+\frac{((3-9t)\gamma^2+(2-4t)\gamma+23t-5)}{2t(1-\gamma)(2\gamma+3)}x^2$
$27 : [123345]$			
$78 : [122345]$			
$152 : [112345]$	$\frac{2(3\gamma+7)}{(2\gamma+3)}$	$-\frac{(3\gamma+1)}{(2\gamma+3)}x$	$+\frac{(-12t\gamma^2-6t\gamma+23t-5)}{2t(1-\gamma)(2\gamma+3)}x^2$
$3 : [123454]$	$\frac{2(3\gamma+7)}{(2\gamma+3)}$	$-\frac{2(\gamma+3)}{(2\gamma+3)}x$	$+\frac{((3-9t)\gamma^2+(2-6t)\gamma+25t-5)}{2t(1-\gamma)(2\gamma+3)}x^2$
$4 : [123453]$			
$5 : [123452]$			
$12 : [123435]$	$\frac{2(3\gamma+7)}{(2\gamma+3)}$	$-\frac{(\gamma+7)}{(2\gamma+3)}x$	$+\frac{((3-7t)\gamma^2+(2-6t)\gamma+23t-5)}{2t(1-\gamma)(2\gamma+3)}x^2$
$17 : [123425]$			
$44 : [123245]$			
$6 : [123451]$	$\frac{2(3\gamma+7)}{(2\gamma+3)}$	$-\frac{3(\gamma+2)}{(2\gamma+3)}x$	$+\frac{(-12t\gamma^2-8t\gamma+25t-5)}{2t(1-\gamma)(2\gamma+3)}x^2$
$22 : [123415]$	$\frac{2(3\gamma+7)}{(2\gamma+3)}$	$-\frac{(2\gamma+7)}{(2\gamma+3)}x$	$+\frac{(-10t\gamma^2-8t\gamma+23t-5)}{2t(1-\gamma)(2\gamma+3)}x^2$
$61 : [123145]$			
$115 : [121345]$			
Group B: $R_B = 2(\gamma+3)/(\gamma+1)$			
$28 : [123344]$	$\frac{(\gamma^2+10\gamma+13)}{((\gamma+1)(\gamma+3))}$	$+\frac{(4-\gamma^2+\gamma)}{(\gamma+1)(\gamma+3)}x$	$+\frac{(-(t-1)\gamma^3+(5-5t)\gamma^2+(15t-1)\gamma+23t-5)}{(2t(1-\gamma)(\gamma+1)(\gamma+3))}x^2$
$79 : [122344]$	$\frac{(\gamma^2+10\gamma+13)}{((\gamma+1)(\gamma+3))}$	$+\frac{4}{(\gamma+1)(\gamma+3)}x$	$+\frac{(-(t-1)\gamma^3+(5-3t)\gamma^2+(13t-1)\gamma+23t-5)}{(2t(1-\gamma)(\gamma+1)(\gamma+3))}x^2$

Table B.2 continues on the next page ...

B.2 Sequence Length $p = 6$

... continued from the previous page.

$l : [\ldots]$	$h_l = c_{11}(l)$	$+2c_{12}(l)x$	$+c_{22}(l)x^2$
$83 : [122334]$	$\frac{(\gamma^2+10\gamma+13)}{((\gamma+1)(\gamma+3))}$	$+\frac{1}{(\gamma+1)}x$	$+\frac{(-(t-1)\gamma^3+(5-3t)\gamma^2+(15t-1)\gamma+21t-5)}{(2t(1-\gamma)(\gamma+1)(\gamma+3))}x^2$
$153 : [112344]$	$\frac{(\gamma^2+10\gamma+13)}{((\gamma+1)(\gamma+3))}$	$+\frac{(4-\gamma^2-\gamma)}{(\gamma+1)(\gamma+3)}x$	$+\frac{(-2t\gamma^3-8t\gamma^2+(11t-3)\gamma+23t-5)}{(2t(1-\gamma)(\gamma+1)(\gamma+3))}x^2$
$169 : [112234]$	$\frac{(\gamma^2+10\gamma+13)}{((\gamma+1)(\gamma+3))}$	$+\frac{(3-\gamma^2)}{(\gamma+1)(\gamma+3)}x$	$+\frac{(-2t\gamma^3-8t\gamma^2+(13t-3)\gamma+21t-5)}{(2t(1-\gamma)(\gamma+1)(\gamma+3))}x^2$
$157 : [112334]$	$\frac{(\gamma^2+10\gamma+13)}{((\gamma+1)(\gamma+3))}$	$+\frac{(3-\gamma)}{(\gamma+1)(\gamma+3)}x$	$+\frac{(-2t\gamma^3-6t\gamma^2+(11t-3)\gamma+21t-5)}{(2t(1-\gamma)(\gamma+1)(\gamma+3))}x^2$
$36 : [123324]$	$\frac{(\gamma^2+10\gamma+13)}{((\gamma+1)(\gamma+3))}$	$-\frac{1}{(\gamma+1)}x$	$+\frac{(-(t-1)\gamma^3+(5-3t)\gamma^2+(15t-1)\gamma+21t-5)}{(2t(1-\gamma)(\gamma+1)(\gamma+3))}x^2$
$10 : [123442]$ $80 : [122343]$	$\frac{(\gamma^2+10\gamma+13)}{((\gamma+1)(\gamma+3))}$	$-\frac{2}{(\gamma+3)}x$	$+\frac{(-(t-1)\gamma^3+(5-5t)\gamma^2+(15t-1)\gamma+23t-5)}{(2t(1-\gamma)(\gamma+1)(\gamma+3))}x^2$
$30 : [123342]$ $45 : [123244]$	$\frac{(\gamma^2+10\gamma+13)}{((\gamma+1)(\gamma+3))}$	$-\frac{2}{(\gamma+3)}x$	$+\frac{(-(t-1)\gamma^3+(5-3t)\gamma^2+(13t-1)\gamma+23t-5)}{(2t(1-\gamma)(\gamma+1)(\gamma+3))}x^2$
$11 : [123441]$ $155 : [112342]$	$\frac{(\gamma^2+10\gamma+13)}{((\gamma+1)(\gamma+3))}$	$-\frac{(\gamma+2)}{(\gamma+3)}x$	$+\frac{(-2t\gamma^3-10t\gamma^2+(13t-3)\gamma+23t-5)}{(2t(1-\gamma)(\gamma+1)(\gamma+3))}x^2$
$62 : [123144]$ $82 : [122341]$	$\frac{(\gamma^2+10\gamma+13)}{((\gamma+1)(\gamma+3))}$	$-\frac{(\gamma+2)}{(\gamma+3)}x$	$+\frac{(-2t\gamma^3-8t\gamma^2+(11t-3)\gamma+23t-5)}{(2t(1-\gamma)(\gamma+1)(\gamma+3))}x^2$
$40 : [123314]$ $161 : [112324]$	$\frac{(\gamma^2+10\gamma+13)}{((\gamma+1)(\gamma+3))}$	$-\frac{3}{(\gamma+3)}x$	$+\frac{(-8t\gamma^2+(11t-3)\gamma+21t-5)}{(2t(1-\gamma)(\gamma+1)(\gamma+3))}x^2$
$91 : [122314]$ $120 : [121334]$	$\frac{(\gamma^2+10\gamma+13)}{((\gamma+1)(\gamma+3))}$	$-\frac{3}{(\gamma+3)}x$	$+\frac{(-2t\gamma^3-6t\gamma^2+(11t-3)\gamma+21t-5)}{(2t(1-\gamma)(\gamma+1)(\gamma+3))}x^2$
$9 : [123443]$	$\frac{(\gamma^2+10\gamma+13)}{((\gamma+1)(\gamma+3))}$	$-\frac{(\gamma^2+\gamma+2)}{(\gamma+1)(\gamma+3)}x$	$+\frac{(-(t-1)\gamma^3+(5-5t)\gamma^2+(15t-1)\gamma+23t-5)}{(2t(1-\gamma)(\gamma+1)(\gamma+3))}x^2$
$31 : [123341]$	$\frac{(\gamma^2+10\gamma+13)}{((\gamma+1)(\gamma+3))}$	$-\frac{(4\gamma+2)}{(\gamma+1)(\gamma+3)}x$	$+\frac{(-2t\gamma^3-8t\gamma^2+(11t-3)\gamma+23t-5)}{(2t(1-\gamma)(\gamma+1)(\gamma+3))}x^2$

Table B.2 continues on the next page ...

Appendix B : Equivalence Classes of Treatments

... continued from the previous page.

$l : [...]$	$h_l = c_{11}(l)$	$+2c_{12}(l)x$	$+c_{22}(l)x^2$
$154 : [112343]$			
$116 : [121344]$	$\frac{(\gamma^2+10\gamma+13)}{((\gamma+1)(\gamma+3))}$	$-\frac{(4\gamma+2)}{(\gamma+1)(\gamma+3)}x$	$+\frac{(-8t\gamma^2+(9t-3)\gamma+23t-5)}{(2t(1-\gamma)(\gamma+1)(\gamma+3))}x^2$
$105 : [122134]$	$\frac{(\gamma^2+10\gamma+13)}{((\gamma+1)(\gamma+3))}$	$-\frac{(\gamma^2+2\gamma+3)}{(\gamma+1)(\gamma+3)}x$	$+\frac{(-2t\gamma^3-8t\gamma^2+(13t-3)\gamma+21t-5)}{(2t(1-\gamma)(\gamma+1)(\gamma+3))}x^2$
$13 : [123434]$	$\frac{(\gamma^2+10\gamma+13)}{((\gamma+1)(\gamma+3))}$	$-\frac{(\gamma^2+3\gamma+8)}{(\gamma+1)(\gamma+3)}x$	$+\frac{(-(t-1)\gamma^3+(5-9t)\gamma^2+(3t-1)\gamma+23t-5)}{(2t(1-\gamma)(\gamma+1)(\gamma+3))}x^2$
$15 : [123432]$	$\frac{(\gamma^2+10\gamma+13)}{((\gamma+1)(\gamma+3))}$	$-\frac{(4\gamma+8)}{(\gamma+1)(\gamma+3)}x$	$+\frac{(-(t-1)\gamma^3+(5-3t)\gamma^2+(13t-1)\gamma+23t-5)}{(2t(1-\gamma)(\gamma+1)(\gamma+3))}x^2$
$18 : [123424]$			
$46 : [123243]$			
$19 : [123423]$	$\frac{(\gamma^2+10\gamma+13)}{((\gamma+1)(\gamma+3))}$	$-\frac{(4\gamma+8)}{(\gamma+1)(\gamma+3)}x$	$+\frac{(-(t-1)\gamma^3+(5-9t)\gamma^2+(3t-1)\gamma+23t-5)}{(2t(1-\gamma)(\gamma+1)(\gamma+3))}x^2$
$49 : [123234]$	$\frac{(\gamma^2+10\gamma+13)}{((\gamma+1)(\gamma+3))}$	$-\frac{3}{(\gamma+1)}x$	$+\frac{(-(t-1)\gamma^3+(5-7t)\gamma^2+(3t-1)\gamma+21t-5)}{(2t(1-\gamma)(\gamma+1)(\gamma+3))}x^2$
$21 : [123421]$	$\frac{(\gamma^2+10\gamma+13)}{((\gamma+1)(\gamma+3))}$	$-\frac{(\gamma^2+5\gamma+8)}{(\gamma+1)(\gamma+3)}x$	$+\frac{(-2t\gamma^3-8t\gamma^2+(11t-3)\gamma+23t-5)}{(2t(1-\gamma)(\gamma+1)(\gamma+3))}x^2$
$23 : [123414]$			
$118 : [121342]$			
$25 : [123412]$	$\frac{(\gamma^2+10\gamma+13)}{((\gamma+1)(\gamma+3))}$	$-\frac{(\gamma^2+5\gamma+8)}{(\gamma+1)(\gamma+3)}x$	$+\frac{(-2t\gamma^3-14t\gamma^2+(t-3)\gamma+23t-5)}{(2t(1-\gamma)(\gamma+1)(\gamma+3))}x^2$
$16 : [123431]$	$\frac{(\gamma^2+10\gamma+13)}{((\gamma+1)(\gamma+3))}$	$-\frac{(6\gamma+8)}{(\gamma+1)(\gamma+3)}x$	$+\frac{(-2t\gamma^3-8t\gamma^2+(11t-3)\gamma+23t-5)}{(2t(1-\gamma)(\gamma+1)(\gamma+3))}x^2$
$24 : [123413]$			
$64 : [123142]$			
$117 : [121343]$			
$48 : [123241]$	$\frac{(\gamma^2+10\gamma+13)}{((\gamma+1)(\gamma+3))}$	$-\frac{(6\gamma+8)}{(\gamma+1)(\gamma+3)}x$	$+\frac{(-8t\gamma^2+(9t-3)\gamma+23t-5)}{(2t(1-\gamma)(\gamma+1)(\gamma+3))}x^2$
$63 : [123143]$			
$132 : [121234]$	$\frac{(\gamma^2+10\gamma+13)}{((\gamma+1)(\gamma+3))}$	$-\frac{(\gamma^2+4\gamma+9)}{(\gamma+1)(\gamma+3)}x$	$+\frac{(-2t\gamma^3-12t\gamma^2+(t-3)\gamma+21t-5)}{(2t(1-\gamma)(\gamma+1)(\gamma+3))}x^2$

Table B.2 continues on the next page ...

... continued from the previous page.

$l : [...]$	$h_l = c_{11}(l)$	$+2c_{12}(l)x$	$+c_{22}(l)x^2$
57 : [123214]	$\frac{(\gamma^2+10\gamma+13)}{((\gamma+1)(\gamma+3))}$	$-\frac{(5\gamma+9)}{(\gamma+1)(\gamma+3)}x$	$+\frac{(-2t\gamma^3-6t\gamma^2+(11t-3)\gamma+21t-5)}{(2t(1-\gamma)(\gamma+1)(\gamma+3))}x^2$
66 : [123134]			
124 : [121324]			
70 : [123124]	$\frac{(\gamma^2+10\gamma+13)}{((\gamma+1)(\gamma+3))}$	$-\frac{(5\gamma+9)}{(\gamma+1)(\gamma+3)}x$	$+\frac{(-12t\gamma^2-(t+3)\gamma+21t-5)}{(2t(1-\gamma)(\gamma+1)(\gamma+3))}x^2$

Group C: $R_C = 6/(\gamma+1)$

170 : [112233]	$\frac{4}{(\gamma+1)}$	$+\frac{8}{3(\gamma+1)}x$	$+\frac{((9t-1)\gamma+21t-5)}{6t(1-\gamma)(\gamma+1)}x^2$
86 : [122331]	$\frac{4}{(\gamma+1)}$	$+\frac{2}{3(\gamma+1)}x$	$+\frac{((9t-1)\gamma+21t-5)}{6t(1-\gamma)(\gamma+1)}x^2$
106 : [122133]			
159 : [112332]			
39 : [123321]	$\frac{4}{(\gamma+1)}$	$-\frac{4}{3(\gamma+1)}x$	$+\frac{((9t-1)\gamma+21t-5)}{6t(1-\gamma)(\gamma+1)}x^2$
92 : [122313]			
122 : [121332]			
42 : [123312]	$\frac{4}{(\gamma+1)}$	$-\frac{4}{3(\gamma+1)}x$	$+\frac{(-(3t+1)\gamma+21t-5)}{6t(1-\gamma)(\gamma+1)}x^2$
133 : [121233]			
162 : [112323]			
58 : [123213]	$\frac{4}{(\gamma+1)}$	$-\frac{10}{3(\gamma+1)}x$	$+\frac{((9t-1)\gamma+21t-5)}{6t(1-\gamma)(\gamma+1)}x^2$
68 : [123132]			
125 : [121323]			
71 : [123123]	$\frac{4}{(\gamma+1)}$	$-\frac{10}{3(\gamma+1)}x$	$+\frac{(-(15t+1)\gamma+21t-5)}{6t(1-\gamma)(\gamma+1)}x^2$
52 : [123231]	$\frac{4}{(\gamma+1)}$	$-\frac{10}{3(\gamma+1)}x$	$+\frac{(-(3t+1)\gamma+21t-5)}{6t(1-\gamma)(\gamma+1)}x^2$

Group D: $R_D = 6(\gamma+1)/(2\gamma+1)$

8 : [123444]	$\frac{(2\gamma+4)}{(\gamma+1)}$	$-\frac{(4\gamma-3)}{3(\gamma+1)}x$	$+\frac{((4-8t)\gamma^2+(t+1)\gamma+23t-5)}{6t(1-\gamma)(\gamma+1)}x^2$

Table B.2 continues on the next page ...

Appendix B : Equivalence Classes of Treatments

... continued from the previous page.

$l : [...]$	$h_l = c_{11}(l)$	$+2c_{12}(l)x$	$+c_{22}(l)x^2$
32 : [123334]	$\frac{(2\gamma+4)}{(\gamma+1)}$	$-\frac{(2\gamma-1)}{3(\gamma+1)}x$	$+\frac{((4-8t)\gamma^2+(5t+1)\gamma+19t-5)}{6t(1-\gamma)(\gamma+1)}x^2$
95 : [122234]			
189 : [111234]	$\frac{(2\gamma+4)}{(\gamma+1)}$	$-\frac{(4\gamma-1)}{3(\gamma+1)}x$	$+\frac{(-12t\gamma^2+(t-3)\gamma+19t-5)}{6t(1-\gamma)(\gamma+1)}x^2$
14 : [123433]	$\frac{(2\gamma+4)}{(\gamma+1)}$	$-\frac{(2\gamma+3)}{3(\gamma+1)}x$	$+\frac{((4-8t)\gamma^2+(9t+1)\gamma+23t-5)}{6t(1-\gamma)(\gamma+1)}x^2$
20 : [123422]			
29 : [123343]			
81 : [122342]			
53 : [123224]	$\frac{(2\gamma+4)}{(\gamma+1)}$	$-\frac{5}{3(\gamma+1)}x$	$+\frac{((4-4t)\gamma^2+(9t+1)\gamma+19t-5)}{6t(1-\gamma)(\gamma+1)}x^2$
87 : [122324]			
26 : [123411]	$\frac{(2\gamma+4)}{(\gamma+1)}$	$-\frac{(4\gamma+3)}{3(\gamma+1)}x$	$+\frac{(-12t\gamma^2+(5t-3)\gamma+23t-5)}{6t(1-\gamma)(\gamma+1)}x^2$
156 : [112341]			
74 : [123114]	$\frac{(2\gamma+4)}{(\gamma+1)}$	$-\frac{(2\gamma+5)}{3(\gamma+1)}x$	$+\frac{(-8t\gamma^2+(5t-3)\gamma+19t-5)}{6t(1-\gamma)(\gamma+1)}x^2$
142 : [121134]			
165 : [112314]			
179 : [112134]			
47 : [123242]	$\frac{(2\gamma+4)}{(\gamma+1)}$	$-\frac{3}{(\gamma+1)}x$	$+\frac{((4-4t)\gamma^2+(5t+1)\gamma+23t-5)}{6t(1-\gamma)(\gamma+1)}x^2$
65 : [123141]	$\frac{(2\gamma+4)}{(\gamma+1)}$	$-\frac{(2\gamma+9)}{3(\gamma+1)}x$	$+\frac{(-8t\gamma^2+(t-3)\gamma+23t-5)}{6t(1-\gamma)(\gamma+1)}x^2$
119 : [121341]			
128 : [121314]	$\frac{(2\gamma+4)}{(\gamma+1)}$	$-\frac{11}{3(\gamma+1)}x$	$+\frac{(-3(t+1)\gamma+19t-5)}{6t(1-\gamma)(\gamma+1)}x^2$

Group E: $R_E = 2(\gamma^2+5\gamma+3)/((\gamma+1)(2\gamma+1))$

84 : [122333]	$\frac{(7\gamma+11)}{(\gamma^2+5\gamma+3)}$	$+\frac{(2\gamma+7)}{(\gamma^2+5\gamma+3)}x$	$+\frac{((t+7)\gamma^2+(20t-2)\gamma+21t-5)}{2t(1-\gamma)(\gamma^2+5\gamma+3)}x^2$
96 : [122233]	$\frac{(7\gamma+11)}{(\gamma^2+5\gamma+3)}$	$+\frac{(3\gamma+6)}{(\gamma^2+5\gamma+3)}x$	$+\frac{((t+7)\gamma^2+(22t-2)\gamma+19t-5)}{2t(1-\gamma)(\gamma^2+5\gamma+3)}x^2$

Table B.2 continues on the next page ...

... continued from the previous page.

$l : [...]$	$h_l = c_{11}(l)$	$+2c_{12}(l)x$	$+c_{22}(l)x^2$
158 : [112333]	$\frac{(7\gamma+11)}{(\gamma^2+5\gamma+3)}$	$+\frac{(-\gamma+7)}{(\gamma^2+5\gamma+3)}x$	$+\frac{(-2t\gamma^2+(14t-4)\gamma+21t-5)}{2t(1-\gamma)(\gamma^2+5\gamma+3)}x^2$
173 : [112223]	$\frac{(7\gamma+11)}{(\gamma^2+5\gamma+3)}$	$+\frac{(\gamma+5)}{(\gamma^2+5\gamma+3)}x$	$+\frac{(-2t\gamma^2+(18t-4)\gamma+17t-5)}{2t(1-\gamma)(\gamma^2+5\gamma+3)}x^2$
190 : [111233]	$\frac{(7\gamma+11)}{(\gamma^2+5\gamma+3)}$	$+\frac{(-\gamma+6)}{(\gamma^2+5\gamma+3)}x$	$+\frac{(-(3t+1)\gamma^2+(14t-6)\gamma+19t-5)}{2t(1-\gamma)(\gamma^2+5\gamma+3)}x^2$
193 : [111223]	$\frac{(7\gamma+11)}{(\gamma^2+5\gamma+3)}$	$+\frac{5}{(\gamma^2+5\gamma+3)}x$	$+\frac{(-(3t+1)\gamma^2+(16t-6)\gamma+17t-5)}{2t(1-\gamma)(\gamma^2+5\gamma+3)}x^2$
38 : [123322] 85 : [122332]	$\frac{(7\gamma+11)}{(\gamma^2+5\gamma+3)}$	$+\frac{(2\gamma+1)}{(\gamma^2+5\gamma+3)}x$	$+\frac{((3t+7)\gamma^2+(30t-2)\gamma+21t-5)}{2t(1-\gamma)(\gamma^2+5\gamma+3)}x^2$
163 : [112322] 171 : [112232]	$\frac{(7\gamma+11)}{(\gamma^2+5\gamma+3)}$	$+\frac{(-\gamma+1)}{(\gamma^2+5\gamma+3)}x$	$+\frac{(-2t\gamma^2+(26t-4)\gamma+21t-5)}{2t(1-\gamma)(\gamma^2+5\gamma+3)}x^2$
34 : [123332]	$\frac{(7\gamma+11)}{(\gamma^2+5\gamma+3)}$	$+0$	$+\frac{((t+7)\gamma^2+(22t-2)\gamma+19t-5)}{2t(1-\gamma)(\gamma^2+5\gamma+3)}x^2$
143 : [121133] 180 : [112133]	$\frac{(7\gamma+11)}{(\gamma^2+5\gamma+3)}$	$-\frac{\gamma}{(\gamma^2+5\gamma+3)}x$	$+\frac{((t-1)\gamma^2+(22t-6)\gamma+19t-5)}{2t(1-\gamma)(\gamma^2+5\gamma+3)}x^2$
112 : [122113] 176 : [112213]	$\frac{(7\gamma+11)}{(\gamma^2+5\gamma+3)}$	$-\frac{1}{(\gamma^2+5\gamma+3)}x$	$+\frac{((t-1)\gamma^2+(24t-6)\gamma+17t-5)}{2t(1-\gamma)(\gamma^2+5\gamma+3)}x^2$
43 : [123311] 160 : [112331]	$\frac{(7\gamma+11)}{(\gamma^2+5\gamma+3)}$	$-\frac{(2\gamma-1)}{(\gamma^2+5\gamma+3)}x$	$+\frac{(-(3t+1)\gamma^2+(24t-6)\gamma+21t-5)}{2t(1-\gamma)(\gamma^2+5\gamma+3)}x^2$
94 : [122311] 172 : [112231]	$\frac{(7\gamma+11)}{(\gamma^2+5\gamma+3)}$	$-\frac{(2\gamma-1)}{(\gamma^2+5\gamma+3)}x$	$+\frac{(-(t+1)\gamma^2+(22t-6)\gamma+21t-5)}{2t(1-\gamma)(\gamma^2+5\gamma+3)}x^2$
121 : [121333]	$\frac{(7\gamma+11)}{(\gamma^2+5\gamma+3)}$	$-\frac{(4\gamma-1)}{(\gamma^2+5\gamma+3)}x$	$+\frac{(-4t\gamma^2+(16t-4)\gamma+21t-5)}{2t(1-\gamma)(\gamma^2+5\gamma+3)}x^2$

Table B.2 continues on the next page ...

Appendix B : Equivalence Classes of Treatments

... continued from the previous page.

$l : [...]$	$h_l = c_{11}(l)$	$+2c_{12}(l)x$	$+c_{22}(l)x^2$
35 : [123331]	$\frac{(7\gamma+11)}{(\gamma^2+5\gamma+3)}$	$-\frac{3\gamma}{(\gamma^2+5\gamma+3)}x$	$+\frac{(-6t\gamma^2+(20t-4)\gamma+19t-5)}{2t(1-\gamma)(\gamma^2+5\gamma+3)}x^2$
98 : [122231]	$\frac{(7\gamma+11)}{(\gamma^2+5\gamma+3)}$	$-\frac{3\gamma}{(\gamma^2+5\gamma+3)}x$	$+\frac{(-2t\gamma^2+(16t-4)\gamma+19t-5)}{2t(1-\gamma)(\gamma^2+5\gamma+3)}x^2$
191 : [111232]	$\frac{(7\gamma+11)}{(\gamma^2+5\gamma+3)}$	$-\frac{4\gamma}{(\gamma^2+5\gamma+3)}x$	$+\frac{(-(7t+1)\gamma^2+(18t-6)\gamma+19t-5)}{2t(1-\gamma)(\gamma^2+5\gamma+3)}x^2$
102 : [122213]	$\frac{(7\gamma+11)}{(\gamma^2+5\gamma+3)}$	$-\frac{(2\gamma+1)}{(\gamma^2+5\gamma+3)}x$	$+\frac{(-4t\gamma^2+(20t-4)\gamma+17t-5)}{2t(1-\gamma)(\gamma^2+5\gamma+3)}x^2$
37 : [123323] 50 : [123233]	$\frac{(7\gamma+11)}{(\gamma^2+5\gamma+3)}$	$-\frac{(\gamma+5)}{(\gamma^2+5\gamma+3)}x$	$+\frac{((t+7)\gamma^2+(20t-2)\gamma+21t-5)}{2t(1-\gamma)(\gamma^2+5\gamma+3)}x^2$
164 : [112321]	$\frac{(7\gamma+11)}{(\gamma^2+5\gamma+3)}$	$-\frac{(5\gamma+2)}{(\gamma^2+5\gamma+3)}x$	$+\frac{(-(3t+1)\gamma^2+(24t-6)\gamma+21t-5)}{2t(1-\gamma)(\gamma^2+5\gamma+3)}x^2$
54 : [123223] 88 : [122323]	$\frac{(7\gamma+11)}{(\gamma^2+5\gamma+3)}$	$-\frac{6}{(\gamma^2+5\gamma+3)}x$	$+\frac{((7-t)\gamma^2+(18t-2)\gamma+19t-5)}{2t(1-\gamma)(\gamma^2+5\gamma+3)}x^2$
108 : [122131] 123 : [121331]	$\frac{(7\gamma+11)}{(\gamma^2+5\gamma+3)}$	$-\frac{(2\gamma+5)}{(\gamma^2+5\gamma+3)}x$	$+\frac{(-(t+1)\gamma^2+(22t-6)\gamma+21t-5)}{2t(1-\gamma)(\gamma^2+5\gamma+3)}x^2$
41 : [123313] 67 : [123133] 107 : [122132] 126 : [121322]	$\frac{(7\gamma+11)}{(\gamma^2+5\gamma+3)}$	$-\frac{(4\gamma+5)}{(\gamma^2+5\gamma+3)}x$	$+\frac{(-2t\gamma^2+(26t-4)\gamma+21t-5)}{2t(1-\gamma)(\gamma^2+5\gamma+3)}x^2$
72 : [123122] 93 : [122312]	$\frac{(7\gamma+11)}{(\gamma^2+5\gamma+3)}$	$-\frac{(4\gamma+5)}{(\gamma^2+5\gamma+3)}x$	$+\frac{(-4t\gamma^2+(16t-4)\gamma+21t-5)}{2t(1-\gamma)(\gamma^2+5\gamma+3)}x^2$
56 : [123221] 90 : [122321]	$\frac{(7\gamma+11)}{(\gamma^2+5\gamma+3)}$	$-\frac{(3\gamma+6)}{(\gamma^2+5\gamma+3)}x$	$+\frac{(2t\gamma^2+(24t-4)\gamma+19t-5)}{2t(1-\gamma)(\gamma^2+5\gamma+3)}x^2$
109 : [122123]	$\frac{(7\gamma+11)}{(\gamma^2+5\gamma+3)}$	$-\frac{(2\gamma+7)}{(\gamma^2+5\gamma+3)}x$	$+\frac{((16t-4)\gamma+17t-5)}{2t(1-\gamma)(\gamma^2+5\gamma+3)}x^2$

Table B.2 continues on the next page ...

... continued from the previous page.

$l : [...]$	$h_l = c_{11}(l)$	$+2c_{12}(l)x$	$+c_{22}(l)x^2$
$136 : [121223]$			
$60 : [123211]$	$\frac{(7\gamma+11)}{(\gamma^2+5\gamma+3)}$	$-\frac{(5\gamma+5)}{(\gamma^2+5\gamma+3)}x$	$+\frac{(-(3t+1)\gamma^2+(24t-6)\gamma+21t-5)}{2t(1-\gamma)(\gamma^2+5\gamma+3)}x^2$
$75 : [123113]$	$\frac{(7\gamma+11)}{(\gamma^2+5\gamma+3)}$	$-\frac{(4\gamma+6)}{(\gamma^2+5\gamma+3)}x$	$+\frac{((t-1)\gamma^2+(22t-6)\gamma+19t-5)}{2t(1-\gamma)(\gamma^2+5\gamma+3)}x^2$
$144 : [121132]$			
$181 : [112132]$			
$166 : [112313]$			
$76 : [123112]$	$\frac{(7\gamma+11)}{(\gamma^2+5\gamma+3)}$	$-\frac{(4\gamma+6)}{(\gamma^2+5\gamma+3)}x$	$+\frac{(-(9t+1)\gamma^2+(14t-6)\gamma+19t-5)}{2t(1-\gamma)(\gamma^2+5\gamma+3)}x^2$
$167 : [112312]$			
$146 : [121123]$	$\frac{(7\gamma+11)}{(\gamma^2+5\gamma+3)}$	$-\frac{(3\gamma+7)}{(\gamma^2+5\gamma+3)}x$	$+\frac{(-(7t+1)\gamma^2+(14t-6)\gamma+17t-5)}{2t(1-\gamma)(\gamma^2+5\gamma+3)}x^2$
$183 : [112123]$			
$51 : [123232]$	$\frac{(7\gamma+11)}{(\gamma^2+5\gamma+3)}$	$-\frac{(\gamma+11)}{(\gamma^2+5\gamma+3)}x$	$+\frac{((7-3t)\gamma^2+(6t-2)\gamma+21t-5)}{2t(1-\gamma)(\gamma^2+5\gamma+3)}x^2$
$59 : [123212]$	$\frac{(7\gamma+11)}{(\gamma^2+5\gamma+3)}$	$-\frac{(4\gamma+11)}{(\gamma^2+5\gamma+3)}x$	$+\frac{(-2t\gamma^2+(14t-4)\gamma+21t-5)}{2t(1-\gamma)(\gamma^2+5\gamma+3)}x^2$
$134 : [121232]$			
$69 : [123131]$	$\frac{(7\gamma+11)}{(\gamma^2+5\gamma+3)}$	$-\frac{(5\gamma+11)}{(\gamma^2+5\gamma+3)}x$	$+\frac{(-(3t+1)\gamma^2+(12t-6)\gamma+21t-5)}{2t(1-\gamma)(\gamma^2+5\gamma+3)}x^2$
$127 : [121321]$			
$73 : [123121]$	$\frac{(7\gamma+11)}{(\gamma^2+5\gamma+3)}$	$-\frac{(5\gamma+11)}{(\gamma^2+5\gamma+3)}x$	$+\frac{(-(7t+1)\gamma^2+(10t-6)\gamma+21t-5)}{2t(1-\gamma)(\gamma^2+5\gamma+3)}x^2$
$135 : [121231]$			
$129 : [121313]$	$\frac{(7\gamma+11)}{(\gamma^2+5\gamma+3)}$	$-\frac{(4\gamma+12)}{(\gamma^2+5\gamma+3)}x$	$+\frac{(-(t+1)\gamma^2+(6t-6)\gamma+19t-5)}{2t(1-\gamma)(\gamma^2+5\gamma+3)}x^2$
$130 : [121312]$			
$139 : [121213]$	$\frac{(7\gamma+11)}{(\gamma^2+5\gamma+3)}$	$-\frac{(3\gamma+13)}{(\gamma^2+5\gamma+3)}x$	$+\frac{(-(3t+1)\gamma^2+(-2t-6)\gamma+17t-5)}{2t(1-\gamma)(\gamma^2+5\gamma+3)}x^2$

Table B.2 continues on the next page ...

Appendix B : Equivalence Classes of Treatments

... continued from the previous page.

$l : [...]$	$h_l = c_{11}(l)$	$+2c_{12}(l)x$	$+c_{22}(l)x^2$
Group F: $R_F = 6/(2\gamma+1)$			
194 : [111222]	$\frac{3}{(2\gamma+1)}$	$+\frac{3}{(2\gamma+1)}x$	$+\frac{((19t-7)\gamma+17t-5)}{6t(1-\gamma)(2\gamma+1)}x^2$
104 : [122211]	$\frac{3}{(2\gamma+1)}$	$+\frac{1}{(2\gamma+1)}x$	$+\frac{((31t-7)\gamma+17t-5)}{6t(1-\gamma)(2\gamma+1)}x^2$
175 : [112221]			
113 : [122112]	$\frac{3}{(2\gamma+1)}$	$-\frac{1}{(2\gamma+1)}x$	$+\frac{((31t-7)\gamma+17t-5)}{6t(1-\gamma)(2\gamma+1)}x^2$
147 : [121122]			
177 : [112212]			
184 : [112122]			
111 : [122121]	$\frac{3}{(2\gamma+1)}$	$-\frac{3}{(2\gamma+1)}x$	$+\frac{((19t-7)\gamma+17t-5)}{6t(1-\gamma)(2\gamma+1)}x^2$
138 : [121221]			
140 : [121212]	$\frac{3}{(2\gamma+1)}$	$-\frac{5}{(2\gamma+1)}x$	$+\frac{(-(5t+7)\gamma+17t-5)}{6t(1-\gamma)(2\gamma+1)}x^2$
Group G: $R_G = 6(\gamma+1)/(3\gamma+1)$			
33 : [123333]	$\frac{(\gamma+3)}{(\gamma+1)}$	$+\frac{(-3\gamma+4)}{3(\gamma+1)}x$	$+\frac{(-(3t-3)\gamma^2+(2t+2)\gamma+19t-5)}{6t(1-\gamma)(\gamma+1)}x^2$
99 : [122223]	$\frac{(\gamma+3)}{(\gamma+1)}$	$+\frac{1}{3(\gamma+1)}x$	$+\frac{(-(3t-3)\gamma^2+(8t+2)\gamma+13t-5)}{6t(1-\gamma)(\gamma+1)}x^2$
199 : [111123]	$\frac{(\gamma+3)}{(\gamma+1)}$	$-\frac{(3\gamma-1)}{3(\gamma+1)}x$	$+\frac{(-6t\gamma^2+(2t-4)\gamma+13t-5)}{6t(1-\gamma)(\gamma+1)}x^2$
55 : [123222]	$\frac{(\gamma+3)}{(\gamma+1)}$	$-\frac{2}{3(\gamma+1)}x$	$+\frac{(-(3t-3)\gamma^2+(14t+2)\gamma+19t-5)}{6t(1-\gamma)(\gamma+1)}x^2$
89 : [122322]			
97 : [122232]			
77 : [123111]	$\frac{(\gamma+3)}{(\gamma+1)}$	$-\frac{(3\gamma+2)}{3(\gamma+1)}x$	$+\frac{(-6t\gamma^2+(8t-4)\gamma+19t-5)}{6t(1-\gamma)(\gamma+1)}x^2$
168 : [112311]			
192 : [111231]			
149 : [121113]	$\frac{(\gamma+3)}{(\gamma+1)}$	$-\frac{5}{3(\gamma+1)}x$	$+\frac{((8t-4)\gamma+13t-5)}{6t(1-\gamma)(\gamma+1)}x^2$

Table B.2 continues on the next page ...

... continued from the previous page.

$l : [\ldots]$	$h_l = c_{11}(l)$	$+2c_{12}(l)x$	$+c_{22}(l)x^2$
186 : [112113]			
196 : [111213]			
131 : [121311]	$\frac{(\gamma+3)}{(\gamma+1)}$	$-\frac{8}{3(\gamma+1)}x$	$+\frac{((8t-4)\gamma+19t-5)}{6t(1-\gamma)(\gamma+1)}x^2$
145 : [121131]			
182 : [112131]			

Group H: $R_H = 2(5\gamma+3)/((\gamma+1)(3\gamma+1))$

174 : [112222]	$\frac{8}{(5\gamma+3)}$	$+\frac{8}{(5\gamma+3)}x$	$+\frac{((15t-3)\gamma+17t-5)}{2t(1-\gamma)(5\gamma+3)}x^2$
200 : [111122]	$\frac{8}{(5\gamma+3)}$	$+\frac{6}{(5\gamma+3)}x$	$+\frac{((15t-7)\gamma+13t-5)}{2t(1-\gamma)(5\gamma+3)}x^2$
114 : [122111]	$\frac{8}{(5\gamma+3)}$	$+\frac{2}{(5\gamma+3)}x$	$+\frac{((27t-7)\gamma+17t-5)}{2t(1-\gamma)(5\gamma+3)}x^2$
178 : [112211]			
195 : [111221]			
101 : [122221]	$\frac{8}{(5\gamma+3)}$	$+0$	$+\frac{((19t-3)\gamma+13t-5)}{2t(1-\gamma)(5\gamma+3)}x^2$
103 : [122212]	$\frac{8}{(5\gamma+3)}$	$-\frac{4}{(5\gamma+3)}x$	$+\frac{((23t-3)\gamma+17t-5)}{2t(1-\gamma)(5\gamma+3)}x^2$
104 : [122122]			
137 : [121222]			
150 : [121112]	$\frac{8}{(5\gamma+3)}$	$-\frac{6}{(5\gamma+3)}x$	$+\frac{((15t-7)\gamma+13t-5)}{2t(1-\gamma)(5\gamma+3)}x^2$
187 : [112112]			
197 : [111212]			
141 : [121211]	$\frac{8}{(5\gamma+3)}$	$-\frac{10}{(5\gamma+3)}x$	$+\frac{((11t-7)\gamma+17t-5)}{2t(1-\gamma)(5\gamma+3)}x^2$
148 : [121121]			
185 : [112121]			

Group I: $R_I = 2(2\gamma+3)/(4\gamma+1)$

100 : [122222]	$\frac{5}{(2\gamma+3)}$	$+\frac{3}{(2\gamma+3)}x$	$+\frac{((3t+5)\gamma+13t-5)}{2t(1-\gamma)(2\gamma+3)}x^2$

Table B.2 continues on the next page ...

Appendix B : Equivalence Classes of Treatments

... continued from the previous page.

$l : [...]$	$h_l = c_{11}(l)$	$+2c_{12}(l)x$	$+c_{22}(l)x^2$
202 : [111112]	$\frac{5}{(2\gamma+3)}$	$+0$	$+\frac{((3t-3)\gamma+5t-5)}{2t(1-\gamma)(2\gamma+3)}x^2$
151 : [121111]	$\frac{5}{(2\gamma+3)}$	$-\frac{3}{(2\gamma+3)}x$	$+\frac{((7t-3)\gamma+13t-5)}{2t(1-\gamma)(2\gamma+3)}x^2$
188 : [112111]			
198 : [111211]			
201 : [111121]			
$R_{203} = 6/(5\gamma+1)$			
203 : [111111]	0	$+0$	$+\frac{5(t-1)}{6t(1-\gamma)}x^2$

Table B.2: All equivalence classes, their representative sequences [...] and h_l functions for sequence length $p = 6$.

B.2.2 Steps of Argumentation

As Proof of Lemma 10. ...

b) Further, observe the expression

$$(h_{28} - h_{170})(x) = \frac{(\gamma^2 + 6\gamma + 1)}{(\gamma+1)(\gamma+3)} - \frac{(3\gamma^2 + 5\gamma + 12)}{3(\gamma+1)(\gamma+3)}x$$
$$+ \frac{-3(t-1)\gamma^3 - 8(3t-2)\gamma^2 - (3t-5)\gamma + 6t}{6t(1-\gamma)(\gamma+1)(\gamma+3)}x^2.$$

Substitution of $x = x_{28}$ gives

$$(h_{28} - h_{170})(x_{28}) = \frac{a(\gamma) - b(\gamma)\sqrt{RT_{x_{28}}}}{12(\gamma+1)[(11t-1)\gamma^3 + 30t\gamma^2 + (10t+1)\gamma - 3t]^2},$$

in which

$a(\gamma) = 60(9t^2 + 10t - 1)\gamma^7 + 2(1465t^2 + 2417t - 156)\gamma^6 + 3(3117t^2 + 3179t - 40)\gamma^5 + 2(5308t^2 + 2375t + 144)\gamma^4 + 12(156t^2 + 205t + 15)\gamma^3 - 8(44t^2 - 512t - 3)\gamma^2 + 3t(639t + 457)\gamma + 774t^2$

and

$b(\gamma) = 60t\gamma^4 - 7(t-13)\gamma^3 + 6(t+28)\gamma^2 + (79t+53)\gamma + 30t.$

The denominator of $(h_{28} - h_{170})(x_{28})$ is positive for all $\gamma \in (0, 1)$ and $t \geq 6$.
The function $a(\gamma) := a_0 + a_1\gamma + \ldots + a_7\gamma^7$ is positive because all coefficients a_i, $i \in \{0, 1, 3, 4, 5, 6, 7\}$, are positive and $(a_2\gamma^2 + a_1\gamma) > (a_2 + a_1)\gamma = (1565t^2 + 5467t + 24)\gamma$ is positive for $\gamma \in (0, 1)$ and $t \geq 6$.
Abbreviate $b(\gamma) = b_0 + b_1\gamma + \ldots + b_4\gamma^4$. All coefficients b_i are positive, except for b_3. Still, $b(\gamma)$ is positive for $\gamma \in (0, 1)$ since $(b_3\gamma^3 + b_1\gamma) > (b_3 + b_1)\gamma = 72(t + 2)\gamma > 0$.
It follows that the numerator of $(h_{28} - h_{170})(x_{28})$ is positive iff

$$a^2(\gamma) - b^2(\gamma)RT_{x_{28}} > 0$$
$$\Leftrightarrow 24[(11t - 1)\gamma^3 + 30t\gamma^2 + (10t + 1)\gamma - 3t]^2 \cdot g(\gamma) > 0$$
$$\Leftrightarrow g(\gamma) := g_0 + g_1\gamma + \ldots + g_8\gamma^8 > 0,$$

in which

$g(\gamma) = 150(5t^2 + 2t + 1)\gamma^8 - 10(49t^2 - 355t - 156)\gamma^7 + (3079t^2 + 7001t + 4956)\gamma^6 + (5929t^2 + 5393t + 4800)\gamma^5 + 6(258t^2 + 887t + 329)\gamma^4 + 4(155t^2 + 1153t + 90)\gamma^3 + (1691t^2 + 1393t + 24)\gamma^2 + (661t + 77)t\gamma + 36t^2$.

It holds that $-10(49t^2 - 355t - 156)\gamma^7 + (3079t^2 + 7001t + 4956)\gamma^6 > (2589t^2 + 10551t + 6516)\gamma^6 > 0$. All other coefficients of γ^i, $0 \leq i \leq 5$, of the function $g(\gamma)$ are positive and it follows that $g(\gamma)$ is positive.
Hence, $h_{28}(x_{28}) > h_{170}(x_{28})$ for all $\gamma \in (0, 1)$ and $t \geq 6$.

c) Analyze

$$(h_{28} - h_8)(x) = \frac{(1 - \gamma)}{(\gamma + 3)} + \frac{(\gamma^2 + 12\gamma + 3)}{3(\gamma + 1)(\gamma + 3)}x$$
$$+ \frac{(5t - 1)\gamma^3 + (8t + 2)\gamma^2 + (19t - 1)\gamma}{6t(1 - \gamma)(\gamma + 1)(\gamma + 3)}x^2$$

and rewrite $h_{28} - h_8$ as $dc_{11} + dc_{12}x + dc_{22}x^2$. The coefficients dc_{11}, dc_{12} and dc_{22} are positive for $\gamma \in (0, 1)$ and $t \geq 6$.
Hence, $h_{28} > h_8$ for all $x \in (0, 1)$.

d) Next to analyze is

$$(h_{28} - h_{14})(x) = \frac{(1 - \gamma)}{(\gamma + 3)} + \frac{(21 + 12\gamma - \gamma^2)}{3(\gamma + 1)(\gamma + 3)}x + \frac{(1 - 5t)\gamma^2 - (5t + 1)\gamma}{6t(\gamma + 1)(\gamma + 3)}x^2.$$

Retype $(h_{28} - h_{14})(x) = dc_{11} + dc_{12}x + dc_{22}x^2$. The coefficients dc_{11} and dc_{12} are positive, dc_{22} is negative for $\gamma \in (0, 1)$ and $t \geq 6$. For $x \in (0, 1)$ we get $dc_{12}x + dc_{22}x^2 >$

Appendix B : Equivalence Classes of Treatments

$(dc_{12} + dc_{22})x^2 = [(1-7t)\gamma^2 + (19t-1)\gamma + 42t]/[6t(\gamma+1)(\gamma+3)] \cdot x^2 > 0$ because $\gamma^2 < \gamma$ and $(1-7t) + (19t-1) > 0$.
Therefore, $h_{28}(x) > h_{14}(x)$ for all $x \in (0,1)$.
Notice, the conclusion of c) and d) is: $h_{28}(x) > h_l(x)$ for $l \in L_D$, $\gamma \in (0,1)$ and $x \in (0,1)$.

e) Next to observe is

$$(h_{28} - h_{84})(x) = \frac{(\gamma^4 + 8\gamma^3 + 27\gamma^2 + 30\gamma + 6)}{(\gamma+1)(\gamma+3)(\gamma^2+5\gamma+3)} - \frac{(\gamma^4 + 6\gamma^3 + 9\gamma^2 + 11\gamma + 9)}{(\gamma+1)(\gamma+3)(\gamma^2+5\gamma+3)} x$$
$$+ \frac{-(t-1)\gamma^5 - (11t-3)\gamma^4 - (37t-1)\gamma^3 - 3(7t+1)\gamma^2 + 2(8t-1)\gamma + 6t}{2t(1-\gamma)(\gamma+1)(\gamma+3)(\gamma^2+5\gamma+3)} x^2.$$

Substitution of $x = x_{28}$ provides

$$(h_{28} - h_{84})(x_{28}) = \frac{a(\gamma) - b(\gamma)\sqrt{RT_{x_{28}}}}{4(\gamma^2+5\gamma+3)[(11t-1)\gamma^3 + 30t\gamma^2 + (10t+1)\gamma - 3t]^2},$$

in which

$a(\gamma) = 20(9t^2 + 10t - 1)\gamma^8 + 2(397t^2 + 359t - 10)\gamma^7 + (4217t^2 - 169t + 48)\gamma^6 + 2(9489t^2 - 856t + 16)\gamma^5 + 2(13292t^2 + 139t - 14)\gamma^4 + 2(1664t^2 + 647t - 6)\gamma^3 - 3t(1347t + 95)\gamma^2 + 36t(119t - 9)\gamma + 972t^2$

and

$b(\gamma) = 20t\gamma^5 + (67t+17)\gamma^4 - 2(52t-7)\gamma^3 - (79t+19)\gamma^2 + 12(15t-1)\gamma + 36t.$

The denominator of $(h_{28} - h_{84})(x_{28})$ is positive for all $\gamma \in (0,1)$ and $t \geq 6$.
Abbreviate $a(\gamma) = a_0 + a_1\gamma + \ldots + a_8\gamma^8$. Just the coefficient a_2 is negative. It follows from $a_2\gamma^2 + a_1\gamma > (a_2 + a_1)\gamma = 3t(81t - 203)\gamma > 0$ that $a(\gamma)$ is positive for $\gamma \in (0,1)$ and $t \geq 6$.
The same approach leads to $b(\gamma) > 0$. Rewrite $b(\gamma) = b_0 + b_1\gamma + \ldots + b_5\gamma^5$. The coefficients b_2 and b_3 are negative, but $\sum_{i=0}^{3} b_i\gamma^i > (\sum_{i=0}^{3} b_i)\gamma = (33t - 17)\gamma$ is positive for all $\gamma \in (0,1)$ and $t \geq 6$.
Thus, $(h_{28} - h_{84})(x_{28}) > 0$ is equivalent to

$$a^2(\gamma) - b^2(\gamma)RT_{x_{28}} > 0$$
$$\Leftrightarrow 8[(11t-1)\gamma^3 + 30t\gamma^2 + (10t+1)\gamma - 3t]^2 \cdot g(\gamma) > 0$$
$$\Leftrightarrow g(\gamma) := g_0 + g_1\gamma + \ldots + g_{10}\gamma^{10} > 0,$$

in which

$$g(\gamma) = 50(5t^2+2t+1)\gamma^{10}+10(127t^2+61t+10)\gamma^9-(431t^2-2041t+90)\gamma^8-$$
$$2(1511t^2-1042t+100)\gamma^7+2(16648t^2-749t+19)\gamma^6+2(37900t^2-$$
$$665t+42)\gamma^5+3(11275t^2+667t+6)\gamma^4-12t(523t+121)\gamma^3+36t(530t-$$
$$59)\gamma^2+432t(25t-1)\gamma+1296t^2.$$

The coefficients g_3, g_7 and g_8 are negative. With $(g_3\gamma^3+g_2\gamma^2) > (g_3+g_2)\gamma^2 = 12t(1067t-298)\gamma^2 > 0$ and $(g_8\gamma^8+g_7\gamma^7+g_6\gamma^6) > (g_8+g_7+g_6)\gamma^6 = (29843t^2+2627t-252)\gamma^6 > 0$ follows that $g(\gamma)$ is positive for all $\gamma \in (0,1)$ and $t \geq 6$.

According to $g(\gamma) > 0$, $h_{28}(x_{28}) > h_{84}(x_{28})$ for all $\gamma \in (0,1)$ and all $t \geq 6$.

f) Now, observe that

$$(h_{28}-h_{38})(x) = \frac{(\gamma^4+8\gamma^3+27\gamma^2+30\gamma+6)}{(\gamma+1)(\gamma+3)(\gamma^2+5\gamma+3)} + \frac{(13\gamma+9-3\gamma^2-6\gamma^3-\gamma^4)}{(\gamma+1)(\gamma+3)(\gamma^2+5\gamma+3)}x$$
$$+\frac{-(t-1)\gamma^5-(13t-3)\gamma^4-(55t-1)\gamma^3-(67t+3)\gamma^2-(14t+2)\gamma+6t}{2t(1-\gamma)(\gamma+1)(\gamma+3)(\gamma^2+5\gamma+3)}x^2.$$

Substitution of $x = x_{28}$ gives

$$(h_{28}-h_{38})(x_{28}) = \frac{(1-\gamma)a(\gamma)-b(\gamma)\sqrt{RT_{x_{28}}}}{4(\gamma^2+5\gamma+3)((11t-1)\gamma^3+30t\gamma^2+(10t+1)\gamma-3t)^2},$$

in which

$$a(\gamma) = -20(9t^2+10t-1)\gamma^7-2(227t^2+483t-20)\gamma^6+(249t^2-1133t-$$
$$8)\gamma^5-(6047t^2-255t+40)\gamma^4-(24123t^2-241t+12)\gamma^3-3t(8013t+$$
$$335)\gamma^2-18t(227t+36)\gamma+1944t^2$$

and

$$b(\gamma) = 20t\gamma^5+(63t+17)\gamma^4-2(151t-13)\gamma^3-(723t+19)\gamma^2-6(35t+4)\gamma+72t.$$

The denominator of $(h_{28}-h_{38})(x_{28})$ is positive for all $\gamma \in (0,1)$ and $t \geq 6$.

The parameter γ is restricted to the interval $(0.95, 1)$ by applying the transformation $\gamma = 0.95 + 0.05\gamma'$, in which $\gamma' \in (0,1)$. The functions $a(\gamma)$ and $b(\gamma)$ alter to

$$a(\gamma') = [-(9t^2+10t-1)\gamma'^7-(1651t^2+2296t-173)\gamma'^6-(115005t^2+208594t-$$
$$11981)\gamma'^5-5(1312939t^2+1936448t-85093)\gamma'^4-5(92433167t^2+$$
$$50050166t-1631767)\gamma'^3-(21099339321t^2+3792553480t-$$
$$79088279)\gamma'^2-(441059980079t^2+35707554774t-276384127)\gamma'-$$
$$3(1056158611135t^2+60474582592t+121356287)]/64000000$$

and

$$b(\gamma') = [t\gamma'^5+(158t+17)\gamma'^4+6(393t+302)\gamma'^3-2(214216t-29431)\gamma'^2-$$
$$(16830847t-548772)\gamma'-(155543238t+609463)]/160000.$$

Appendix B : Equivalence Classes of Treatments

Since all coefficients a_i of γ^i, $0 \leq i \leq 7$, in $a(\gamma')$ are negative for $\gamma' \in (0,1)$ and $t \geq 6$, function $a(\gamma')$ is negative itself.

Rewrite $b(\gamma') = b_5 \gamma'^5 + \ldots + b_1 \gamma' + b_0$. The coefficients b_3, b_4 and b_5 are positive, such that $b(\gamma') < (b_3+b_4+b_5)\gamma'^5 + (b_0+b_1+b_2)\gamma'^5 = -1728 \cdot 10^5 t \gamma'^5 < 0$ for $t \geq 6$ and $\gamma' \in (0,1)$. Both functions, $a(\gamma)$ and $b(\gamma)$, are negative for $\gamma \in [0.95, 1)$. Thus, $(h_{28} - h_{38})(x_{28}) > 0$ is equivalent to

$$b^2(\gamma) RT_{x_{28}} - (1-\gamma)^2 a^2(\gamma) > 0$$
$$\Leftrightarrow 8[(11t-1)\gamma^3 + 30t\gamma^2 + (10t+1)\gamma - 3t]^2 \cdot g(\gamma) > 0$$
$$\Leftrightarrow g(\gamma) > 0,$$

in which

$g(\gamma) = \ 50(5t^2+2t+1)\gamma^9 + 10(124t^2+47t+15)\gamma^8 - (4991t^2-591t-60)\gamma^7 - (31731t^2+2977t+140)\gamma^6 + (8685t^2-10751t-102)\gamma^5 + 3(70811t^2-2379t-6)\gamma^4 + 6t(50151t+1424)\gamma^3 + 72t(1537t+131)\gamma^2 - 432t(21t-4)\gamma - 5184t^2$.

As $a(\gamma)$ and $b(\gamma)$ are negative for $\gamma \in (0.95, 1)$, the function $g(\gamma)$ needs to be restricted to this particular interval. Therefore, apply the transformation $\gamma = 0.95 + 0.05\gamma'$, in which $\gamma' \in (0,1)$. Function $g(\gamma)$ rearranges to

$g(\gamma') = \ [(5t^2 + 2t + 1)\gamma'^9 + (1351t^2 + 530t + 231)\gamma'^8 + 4(25111t^2 + 14824t + 5649)\gamma'^7 - 4(623259t^2 - 801280t - 306019)\gamma'^6 - 2(290528897t^2 - 26095866t - 20112783)\gamma'^5 - 2(7378207115t^2 + 1415158086t - 410555937)\gamma'^4 + 4(1821799859991t^2 - 39444395060t + 2530562861)\gamma'^3 + 4(11238809719909t^2 - 714376954124t + 16876169151)\gamma'^2 + (826994043598981t^2 - 16914338970310t + 162906812649)\gamma' + 5099305337440679t^2 + 19932399226538t - 241396325041]/10240000000$.

Abbreviate $g(\gamma') = g_0 + g_1\gamma' + \ldots + g_9\gamma'^9$. Three coefficients, g_4 g_5 and g_6, are negative. However, $g(\gamma')$ is positive because $g_6\gamma'^6 + g_5\gamma'^5 + g_4\gamma'^4 + g_3\gamma'^3 > (g_6 + g_5 + g_4 + g_3)\gamma'^4 = 8(89172497363t^2 - 20069062445t + 1373101620)\gamma'^4 > 0$ for $t \geq 6$ and $\gamma' \in (0,1)$. It follows that the numerator of $(h_{28} - h_{38})(x_{28})$ is positive for all $\gamma \in (0.95, 1) \supset [\gamma_{\beta(t)}, 1)$, and thus, $h_{28}(x_{28}) > h_{38}(x_{28})$ for all $\gamma \in [\gamma_{\beta(t)}, 1)$ and all $t \geq 6$.

g) Next to analyze is
$$(h_{28}-h_{194})(x) = \frac{2(\gamma^3+9\gamma^2+12\gamma+2)}{(\gamma+1)(\gamma+3)(2\gamma+1)} - \frac{(2\gamma^3+2\gamma^2+3\gamma+5)}{(\gamma+1)(\gamma+3)(2\gamma+1)}x$$
$$+ \frac{-3(t-1)\gamma^4 - 2(13t-10)\gamma^3 - 3(3t-7)\gamma^2 + (29t+4)\gamma + 9t}{3t(1-\gamma)(\gamma+1)(\gamma+3)(2\gamma+1)}x^2.$$

Define function
$$d(\gamma) := \ d_4\gamma^4 + \ldots + d_1\gamma + d_0 = -3(t-1)\gamma^4 - 2(13t-10)\gamma^3 - 3(3t-7)\gamma^2 + (29t+4)\gamma + 9t.$$

The expression $(h_{28} - h_{194})(x)$ is a convex function in x because $d(\gamma) > (d_4 + \ldots + d_1 + d_0)\gamma = 48\gamma > 0$ for all $\gamma \in (0,1)$ and $t \geq 6$. Hence, the second derivative $(h_{28} - h_{194})''(x)$ is positive.

The slope of $(h_{28} - h_{194})(x)$ equals 0 iff $x = x_{st} = \frac{3t(1-\gamma)(2\gamma^3+2\gamma^2+3\gamma+5)}{2 \cdot d(\gamma)}$. The function value of this stationary point must be the minimum of the function and is equal to

$$(h_{28} - h_{194})(x_{st}) = \frac{g(\gamma)}{4(\gamma+1)(\gamma+3)(2\gamma+1)d(\gamma)},$$

in which
$$g(\gamma) = \ -12(t-2)\gamma^7 - 4(103t-94)\gamma^6 - 24(92t-79)\gamma^5 - 8(364t-439)\gamma^4 +$$
$$(871t+2624)\gamma^3 + 3(1097t+240)\gamma^2 + (1313t+64)\gamma + 69t.$$

Abbreviate $g(\gamma)$ as the sum of $g_7\gamma^7 + \ldots + g_1\gamma + g_0$ and reckon that g_7, g_6, g_5 and g_4 are negative coefficients. But since $g(\gamma) > (g_7 + \ldots + g_1 + g_0)\gamma^3 = 9216\gamma^3 > 0$ and $d(\gamma) > 0$ for all $\gamma \in (0,1)$ and $t \geq 6$, $(h_{28} - h_{194})(x_{st})$ is positive as well.

The conclusion of g) is: $h_{28}(x) > h_{194}(x)$ for all $x > 0$ because the minimum of $(h_{28} - h_{194})(x)$ is positive for all $\gamma \in (0,1)$ and all $t \geq 6$.

h) Another expression to examine is
$$(h_{28} - h_{104})(x) = \frac{2(\gamma^3+9\gamma^2+12\gamma+2)}{(\gamma+1)(\gamma+3)(2\gamma+1)} + \frac{(1+5\gamma-2\gamma^3)}{(\gamma+1)(\gamma+3)(2\gamma+1)}x$$
$$+ \frac{-3(t-1)\gamma^4 - 4(8t-5)\gamma^3 - 3(11t-7)\gamma^2 + (11t+4)\gamma + 9t}{3t(1-\gamma)(\gamma+1)(\gamma+3)(2\gamma+1)}x^2.$$

Rewrite $(h_{28} - h_{104})(x) = dc_{11} + dc_{12}x + dc_{22}x^2$. It is clear to detect that dc_{11} and dc_{12} are positive for $\gamma \in (0,1)$ and $t \geq 6$. The sum $dc_{11} + dc_{12} + dc_{22}$ is equivalent to $d(\gamma)/[3t(1-\gamma)(\gamma+1)(\gamma+3)(2\gamma+1)]$, in which

$$d(\gamma) := \ d_4\gamma^4 + \ldots + d_1\gamma + d_0 = 3(t-1)\gamma^4 - 2(11t-10)\gamma^3 - 21\gamma^2 + (61t-4)\gamma + 6t.$$

It holds that $d(\gamma) > d_4\gamma^4 + (d_3+d_2+d_1+d_0)\gamma^2 = (3\gamma^2+45)(t-1)\gamma^2 > 0$ for all $\gamma \in (0,1)$

Appendix B : Equivalence Classes of Treatments

and $t \geq 6$. Hence, $dc_{11} + dc_{12}x + dc_{22}x^2 > (dc_{11} + dc_{12} + dc_{22})x^2$ for all $x \in (0,1)$ and $h_{28}(x) > h_{104}(x)$ for all $x, \gamma \in (0,1)$ and $t \geq 6$.

Notice, summing up g) and h), it follows that $h_{28} > h_l$ for all $l \in L_F$, $x, \gamma \in (0,1)$ and $t \geq 6$.

i) As $-(2t-1)\gamma^2 + 3t > (t+1)\gamma^2$, all coefficients of x^i, $0 \leq i \leq 2$, of

$$(h_{28} - h_{33})(x) = \frac{4}{(\gamma+3)} + \frac{8\gamma}{3(\gamma+1)(\gamma+3)}x + \frac{2[-(2t-1)\gamma^2 + (5t-1)\gamma + 3t]}{3t(1-\gamma)(\gamma+1)(\gamma+3)}x^2,$$

are positive for $x > 0$, $\gamma \in (0,1)$ and $t \geq 6$.

Thus, $h_{28}(x) > h_{33}(x)$ for all $x > 0$, $\gamma \in (0,1)$ and $t \geq 6$.

k) Observe that

$$(h_{28} - h_{55})(x) = \frac{4}{(\gamma+3)} + \frac{(18 + 5\gamma - 3\gamma^2)}{3(\gamma+1)(\gamma+3)}x + \frac{2[-(5t-1)\gamma^2 - (4t+1)\gamma + 3t]}{3t(1-\gamma)(\gamma+1)(\gamma+3)}x^2.$$

Substitution of $x = x_{28}$ gives

$$(h_{28} - h_{55})(x_{28}) = \frac{a(\gamma) - b(\gamma)\sqrt{RT_{x_{28}}}}{2(\gamma+1)((11t-1)\gamma^3 + 30t\gamma^2 + (10t+1)\gamma - 3t)^2},$$

in which

$a(\gamma) = -2t(11t-1)\gamma^7 - (87t^2 - 89t + 8)\gamma^6 + 8(179t^2 - 2t + 1)\gamma^5 + (4471t^2 - 298t + 8)\gamma^4 + (2163t^2 + 254t - 8)\gamma^3 - 37t(76t - 5)\gamma^2 - 3t(395t + 72)\gamma + 648t^2$

and

$b(\gamma) = (11t-1)\gamma^4 - 2(14t-3)\gamma^3 - (104t-3)\gamma^2 - (35t+8)\gamma + 24t.$

The denominator of $(h_{28} - h_{55})(x_{28})$ is positive for all $\gamma \in (0,1)$ and $t \geq 6$.

In order to determine the roots of $(h_{28} - h_{55})(x_{28})$ for $\gamma \in (\gamma_{\beta(t)}, 1)$, apply the transformation $\gamma = 0.95 + 0.05\gamma'$, in which $\gamma' \in (0,1)$ and $\gamma \in (0.95, 1) \supset [\gamma_{\beta(t)}, 1)$. As a result, functions $a(\gamma)$ and $b(\gamma)$ alter to

$a(\gamma') = [-t(11t-1)\gamma'^7 - (2333t^2 - 1023t + 80)\gamma'^6 + (103829t^2 + 105841t - 7520)\gamma'^5 + 5(7548047t^2 + 712683t - 49840)\gamma'^4 + 5(479321563t^2 + 8965487t - 681280)\gamma'^3 + (61472332081t^2 + 444663429t - 13809200)\gamma'^2 + (639250046383t^2 + 5105401827t + 38872480)\gamma' + 3(748654390667t^2 - 1866187657t - 7133360)]/640000000$

and

$b(\gamma') = [(11t-1)\gamma'^4 + 4(69t+11)\gamma'^3 - 2(24847t - 2937)\gamma'^2 - 4(541371t - 21031)\gamma' - (18905109t + 90041)]/160000.$

Abbreviate $a(\gamma') = [a_7\gamma'^7 + \ldots + a_1\gamma' + a_0]/64000000$ and $b(\gamma') = [b_4\gamma'^4 + \ldots + b_1\gamma' + b_0]/160000$.

In order to show that $a(\gamma')$ is positive, observe that all coefficients a_i, $0 \leq i \leq 5$, are negative. Furthermore, $(a_7\gamma'^7 + a_6\gamma'^6 + a_5\gamma'^5) > (a_7 + a_6 + a_5)\gamma'^5 = 5(20297t^2 + 21373t - 1520)\gamma'^5 > 0$. It follows that $a(\gamma')$ is positive for all $\gamma' \in (0,1)$ and $t \geq 6$.

Use the fact that $b_2\gamma'^2 + b_3\gamma'^3 + b_4\gamma'^4 < (b_2 + b_3 + b_4)\gamma'^4 = -(49407t - 5917)\gamma'^4$, b_0 and b_1 are negative for $\gamma' \in (0,1)$ and $t \geq 6$. It follows that the function $b(\gamma')$ is negative for all $\gamma' \in (0,1)$ and all $t \geq 6$.

A positive function $a(\gamma')$ and a negative function $b(\gamma')$ lead to $a(\gamma) - b(\gamma)\sqrt{RT_{x_{28}}} > 0$ for all $\gamma \in (0.95, 1) \supset [\gamma_{\beta(t)}, 1)$.

Hence, $h_{28}(x_{28}) > h_{55}(x_{28})$ for all $\gamma \in [\gamma_{\beta(t)}, 1)$ and all $t \geq 6$.

l) Another expression to analyze is

$$(h_{28} - h_{174})(x) = \frac{(5\gamma^3 + 45\gamma^2 + 63\gamma + 15)}{(\gamma+1)(\gamma+3)(5\gamma+3)} - \frac{(5\gamma^3 + 6\gamma^2 + 9\gamma + 12)}{(\gamma+1)(\gamma+3)(5\gamma+3)}x$$
$$+ \frac{-5(t-1)\gamma^4 - (43t-31)\gamma^3 - (17t-27)\gamma^2 + (47t+1)\gamma + 18t}{2t(1-\gamma)(\gamma+1)(\gamma+3)(5\gamma+3)}x^2.$$

Rewrite $(h_{28} - h_{174})(x) = dc_{11} + dc_{12}x + dc_{22}x^2$. It is clear to observe that dc_{11} is positive and dc_{12} is negative. The coefficient $dc_{22} := \frac{d_4\gamma^4 + \ldots + d_1\gamma + d_0}{2t(1-\gamma)(\gamma+1)(\gamma+3)(5\gamma+3)}$ is positive, because its denominator is positive and $d_0 + d_1\gamma > (d_0 + d_1)\gamma \geq -(d_2 + d_3 + d_4)\gamma^2 > -(d_2\gamma^2 + d_3\gamma^3 + d_4\gamma^4)$. Restrict x to interval $(0,1)$ to get $dc_{11} + dc_{12}x > (dc_{11} + dc_{12})x = \frac{39\gamma^2 + 54\gamma + 3}{(\gamma+1)(\gamma+3)(5\gamma+3)}x > 0$ for all $\gamma \in (0,1)$.

It follows that $h_{28}(x) > h_{174}(x)$ for $x, \gamma \in (0,1)$ and $t \geq 6$.

m) Next, observe the expression

$$(h_{28} - h_{114})(x) = \frac{(5\gamma^3 + 45\gamma^2 + 63\gamma + 15)}{(\gamma+1)(\gamma+3)(5\gamma+3)} + \frac{(6 + 15\gamma - 5\gamma^3)}{(\gamma+1)(\gamma+3)(5\gamma+3)}x$$
$$+ \frac{-5(t-1)\gamma^4 - 5(11t-7)\gamma^3 - (65t-43)\gamma^2 + (11t+13)\gamma + 18t}{2t(1-\gamma)(\gamma+1)(\gamma+3)(5\gamma+3)}x^2.$$

Substitution of $x = x_{28}$ gives

$$(h_{28} - h_{114})(x_{28}) = \frac{a(\gamma) - b(\gamma)\sqrt{RT_{x_{28}}}}{4(5\gamma+3)[(11t-1)\gamma^3 + 30t\gamma^2 + (10t+1)\gamma - 3t]^2},$$

in which

Appendix B : Equivalence Classes of Treatments

$a(\gamma) = 100(9t^2 + 10t - 1)\gamma^7 + 10(771t^2 + 823t - 56)\gamma^6 + (35777t^2 + 17071t - 320)\gamma^5 + 2(28163t^2 + 5552t + 244)\gamma^4 + 2(5837t^2 + 3449t + 210)\gamma^3 - 2(11585t^2 - 4073t - 36)\gamma^2 - 3t(573t - 949)\gamma + 4662t^2$

and

$b(\gamma) = 100t\gamma^4 - 5(65t - 37)\gamma^3 - 2(380t - 163)\gamma^2 - (5t - 113)\gamma + 174t$.

The denominator of $(h_{28} - h_{55})(x_{28})$ is positive for all $\gamma \in (0,1)$ and $t \geq 6$. In order to determine the algebraic signs of $a(\gamma)$ and $b(\gamma)$ for $\gamma \in [\gamma_{\beta(t)}, 1) \subset (0.95, 1)$, apply the transformation $\gamma = 0.95 + 0.05\gamma'$, in which $\gamma' \in (0,1)$ and $\gamma \in (0.95, 1)$. Both functions are modified to

$a(\gamma') = [(9t^2 + 10t - 1)\gamma'^7 + (2739t^2 + 2976t - 245)\gamma'^6 + (387125t^2 + 331738t - 21629)\gamma'^5 + 5(5722371t^2 + 3737808t - 185821)\gamma'^4 + 5(226068607t^2 + 119292998t - 4181431)\gamma'^3 + (23381483241t^2 + 11235076560t - 233563919)\gamma'^2 + (233705284871t^2 + 120293779134t - 940884655)\gamma' + 3(307133962375t^2 + 191881485184t + 398768903)]/12800000$

and

$b(\gamma') = [t\gamma'^4 + (11t + 37)\gamma'^3 - (4579t - 3413)\gamma'^2 - (158879t - 98663)\gamma' - (1142154t - 896287)]/1600$.

The function $a(\gamma')$ is positive for $\gamma' \in (0,1)$, because all coefficients of γ'^i, $0 \leq i \leq 7$, are positive. On the contrary, $b(\gamma')$ is negative for $\gamma' \in (0,1)$. In function $b(\gamma') := [b_4\gamma'^4 + \ldots + b_1\gamma' + b_0]/1600$, just coefficients b_3 and b_4 are positive and, therefore, $b(\gamma') < (b_0 + b_1 + \ldots + b_4)\gamma'^2 = -76800(17t - 13)\gamma'^2 < 0$ for all $\gamma' \in (0,1)$ and $t \geq 6$. Thus, $(h_{28} - h_{114})(x_{28})$ is positive for $\gamma \in (0.95, 1)$, which leads to the conclusion that $h_{28}(x_{28}) > h_{114}(x_{28})$ for $\gamma \in [\gamma_{\beta(t)}, 1)$ and $t \geq 6$.

n) The difference of h_{28} and h_{100} is

$$(h_{28} - h_{100})(x) = \frac{2(\gamma^3 + 9\gamma^2 + 18\gamma + 12)}{(\gamma+1)(\gamma+3)(2\gamma+3)} - \frac{(2\gamma^3 + 4\gamma^2 + \gamma - 3)}{(\gamma+1)(\gamma+3)(2\gamma+3)}x + \frac{-(t-1)\gamma^4 - 4(2t-1)\gamma^3 - (5t+1)\gamma^2 + (15t-4)\gamma + 15t}{t(1-\gamma)(\gamma+1)(\gamma+3)(2\gamma+3)}x^2$$

and positive for all $x, \gamma \in (0,1)$ and $t \geq 6$. Rewrite $(h_{28} - h_{100})(x)$ as $dc_{11} + dc_{12}x + dc_{22}x^2$. Use the fact that $\gamma^i > \gamma^{i+1}$, for all $i \in \mathbb{N}$, the coefficients dc_{11} and dc_{22} are positive. Assume $x \in (0,1)$ to get $dc_{11} + dc_{12}x > (dc_{11} + dc_{12})x = \frac{(14\gamma^2 + 35\gamma + 27)}{(\gamma+1)(\gamma+3)(2\gamma+3)}x > 0$. Hence, $h_{28}(x) > h_{100}(x)$ for all $x, \gamma \in (0,1)$ and $t \geq 6$.

o) The last h_l function to compare to h_{28} is $l = 151$. In this manner,

$$(h_{28} - h_{151})(x) = \frac{2(\gamma^3 + 9\gamma^2 + 18\gamma + 12)}{(\gamma + 1)(\gamma + 3)(2\gamma + 3)} + \frac{(21 + 23\gamma + 2\gamma^2 - 2\gamma^3)}{(\gamma + 1)(\gamma + 3)(2\gamma + 3)}x$$
$$+ \frac{-(t-1)\gamma^4 - 2(5t-4)\gamma^3 - (13t-15)\gamma^2 + (9t+8)\gamma + 15t}{t(1-\gamma)(\gamma+1)(\gamma+3)(2\gamma+3)}x^2.$$

Again, use the fact that $\gamma^i > \gamma^{i+1}$, for all $i \in \mathbb{N}$, and it can be seen that coefficients dc_{11}, dc_{12} and dc_{22} are positive by rewriting $(h_{28} - h_{151})(x) = dc_{11} + dc_{12}x + dc_{22}x^2$.
Thus, $h_{28}(x) > h_{151}(x)$ for all $x > 0$, $\gamma \in (0, 1)$ and $t \geq 6$.
Notice, summing up n) and o), it follows that $h_{28} > h_l$ for all $l \in L_I \cup \{203\}$, $x, \gamma \in (0, 1)$ and $t \geq 6$.

□

As Proof of Lemma 11. ...

c) In this manner, observe that

$$(h_2 - h_{170})(x) = \frac{2(3\gamma^2 + 6\gamma + 1)}{(\gamma + 1)(2\gamma + 3)} - \frac{9\gamma^2 + 25\gamma + 24}{3(\gamma + 1)(2\gamma + 3)}x$$
$$+ \frac{-9(3t-1)\gamma^3 - (63t-17)\gamma^2 - (3t-1)\gamma + 12t}{6t(1-\gamma)(\gamma+1)(2\gamma+3)}x^2.$$

Substitution of $x = x_2$ provides

$$(h_2 - h_{170})(x_2) = \frac{a(\gamma) - 2b(\gamma)\sqrt{RT_{x_2}}}{3\gamma^2(\gamma+1)[(23t-1)\gamma + 7t + 1]^2},$$

with
$a(\gamma) = 45(21t^2 + 22t - 1)\gamma^5 - 2(551t^2 - 599t - 3)\gamma^4 + (2993t^2 - 620t + 33)\gamma^3 + 2(160t^2 + 350t + 3)\gamma^2 - 12t(113t - 36)\gamma + 900t^2$

and
$b(\gamma) = 45t\gamma^3 - (25t - 34)\gamma^2 - 2(10t - 7)\gamma + 30t.$

The denominator of $(h_2 - h_{170})(x_2)$ is positive for all $\gamma \in (0, 1)$ and $t \geq 6$.
The function $a(\gamma)$ is positive for $\gamma \in (0, 1)$. Just the coefficient $-2(551t^2 - 599t - 3)$ of γ^4 is negative for all $t \geq 6$, but as $-2(551t^2 - 599t - 3)\gamma^4 + (2993t^2 - 620t + 33)\gamma^3 >$

Appendix B : Equivalence Classes of Treatments

$(1891t^2 + 578t + 39)\gamma^4 > 0$, the function $a(\gamma)$ is positive for all $t \geq 6$.
Analyze the characteristics of $b(\gamma)$ by taking its derivative twice. The slope, $b'(\gamma)$, reveals one stationary point in interval $(0,1)$ at $\gamma = \gamma_{st} = \frac{1}{135t}[\sqrt{3325t^2 - 3590t + 1156} + 25t - 34]$. A change in the curvature of $b(\gamma)$ in interval $(0,1)$ is located at $\gamma_{inf} = (25t - 34)/135t$. Hence, the function $b(\gamma)$ is concave for all $\gamma < \gamma_{inf}$, and convex for all $\gamma > \gamma_{inf}$. Since $\gamma_{st} > \gamma_{inf}$ and $b(\gamma)$ is convex for all $\gamma > \gamma_{inf}$, γ_{st} is the location of a minimum of $b(\gamma)$. Some simple equation transformations confirm that the function value of

$$b(\gamma_{st}) = \frac{2(5t+4)(140650t^2 - 58055t + 9826) - 2\sqrt{(3325t^2 - 3590t + 1156)^3}}{54675t^2}$$

is positive for all $t \geq 6$. Thus, $b(\gamma)$ is positive for all $\gamma \in (0,1)$ and $t \geq 6$ because the minimum of $b(\gamma)$ of interval $(0,1)$ is positive.

Use that the function $a(\gamma)$ and $b(\gamma)$ are positive. It follows that $(h_2 - h_{170})(x_2) > 0$ is equivalent to

$$a^2(\gamma) - 4b^2(\gamma)RT_{x_2} > 0$$
$$\Leftrightarrow 3\gamma^2[(23t-1)\gamma + 7t + 1]^2 \cdot g(\gamma) > 0$$
$$\Leftrightarrow g(\gamma) := g_6\gamma^6 + \ldots + g_1\gamma + g_0 > 0,$$

in which

$g(\gamma) = \quad 675(5t^2+2t+1)\gamma^6 - 30(263t^2-98t-39)\gamma^5 + (6547t^2-2578t+687)\gamma^4 +$
$\quad\quad 12(274t^2+152t+13)\gamma^3 - 12(427t^2-142t-1)\gamma^2 + 8t(277t+20)\gamma + 288t^2.$

In order to show that $g(\gamma)$ is positive for all $\gamma \in (0,1)$, decompose $g(\gamma)$ into $G_{03}(\gamma) + G_{46}(\gamma)\gamma^4$. Iff $G_{03}(\gamma)$ and $G_{46}(\gamma)$ are both positive for all $\gamma \in (0,1)$, the function $g(\gamma)$ is positive, as well.

c1) The first derivative of $G_{03}(\gamma)$ is presented by $G'_{03}(\gamma) = g_1 + 2g_2\gamma + 3g_3\gamma^3$.
For $t = 6$, $G'_{03}(\gamma) = 388404\gamma^2 - 248456\gamma + 80736 > 0$ for all γ in the interval $(0,1)$. A positive slope $G'_{03}(\gamma)$ means that the function $G_{03}(\gamma)$ is increasing in γ. Thus, $G_{03}(\gamma \searrow 0) = g_0 > 0$ is the local minimum of $G_{03}(\gamma)$ of the interval $(0,1)$. Since the minimum of $G_{03}(\gamma)$ is positive for all $\gamma \in (0,1)$ and $t = 6$, the function $G_{03}(\gamma)$ is positive in the described domains of the parameters γ and t.
For $t > 6$, stationary points of interval $(0,1)$ are derived from $G'_{03}(\gamma) \stackrel{!}{=} 0$, which is fulfilled iff γ equals

$$\gamma_{st1/2} = \frac{\mp\sqrt{30533t^4 - 216436t^3 + 6028t^2 - 236t + 1} + 427t^2 - 142t - 1}{3(274t^2 + 152t + 13)}.$$

The second derivative of $G_{03}(\gamma)$ of interval $(0,1)$ is 0 iff $\gamma = \gamma_{inf} = \frac{427t^2-142t-1}{3(274t^2+152t+13)}$. Thus, $G_{03}(\gamma)$ is concave for all $\gamma < \gamma_{inf}$, and $G_{03}(\gamma)$ is convex for all $\gamma > \gamma_{inf}$. Compare γ_{inf} with $\gamma_{st1/2}$ to get $\gamma_{st1} < \gamma_{inf} < \gamma_{st2}$. The relation implies that $G_{03}(\gamma_{st1})$ is a maximum and $G_{03}(\gamma_{st2})$ is a minimum of $G_{03}(\gamma)$. Therefore, iff $G_{03}(\gamma)$ is positive at its local minima with locations 0 and γ_{st2}, the function $G_{03}(\gamma)$ is positive for all $\gamma \in (0,1)$.

Calculate the function values of both potential minima and observe that $G_{03}(\gamma \searrow 0) = g_0$ is positive. Some simple equation transformations confirm that

$$G_{03}(\gamma_{st2}) = \frac{8[V - \sqrt{(30533t^4 - 216436t^3 + 6028t^2 - 236t + 1)^3}]}{9(274t^2 + 152t + 13)^2},$$

in which $V = 43695479t^6 + 133282614t^5 - 27480582t^4 + 1141174t^3 - 16716t^2 - 354t + 1$ is positive for $t > 6$. It follows that $G_{03}(\gamma)$ is positive for all $\gamma \in (0,1)$ and all $t > 6$.

c2) $G_{46}(\gamma)$ is a convex function of γ because its second derivative, $G_{46}''(\gamma) = 2g_6$, is positive for $t \geq 6$. There exists just one stationary point of $G_{46}(\gamma)$ at $-g_5/(2g_6) > 1$, which must be a minimum. Thus, $G_{46}(\gamma)$ is decreasing in γ in interval $(0,1)$. Its local minima is $G_{46}(\gamma \nearrow 1) = 4(508t^2 + 428t + 633)$, which is positive for all $t \geq 6$ and, therefore, $G_{46}(\gamma)$ is positive in the entire interval $(0,1)$.

We get $g(\gamma) = G_{03}(\gamma) + G_{46}(\gamma)\gamma^4 > 0$ for all $\gamma \in (0,1)$ and $t \geq 6$. The consequence is that $h_2(x_2) > h_{170}(x_2)$ for all $\gamma \in (0,1)$ and $t \geq 6$.

d) Next to analyze is

$$(h_2 - h_{84})(x) = \frac{3(2\gamma^3 + 10\gamma^2 + 15\gamma + 3)}{(2\gamma+3)(\gamma^2+5\gamma+3)} - \frac{3\gamma^3 + 19\gamma^2 + 29\gamma + 21}{(2\gamma+3)(\gamma^2+5\gamma+3)}x$$
$$+ \frac{-3(3t-1)\gamma^4 - (53t-3)\gamma^3 - (75t+3)\gamma^2 + (5t-3)\gamma + 12t}{2t(1-\gamma)(2\gamma+3)(\gamma^2+5\gamma+3)}x^2.$$

Substitution of $x = x_2$ gives

$$(h_2 - h_{84})(x_2) = \frac{a(\gamma) - 2b(\gamma)\sqrt{RT_{x_2}}}{\gamma^2(\gamma^2+5\gamma+3)[(23t-1)\gamma + 7t + 1]^2},$$

with

$a(\gamma) = \ 15(21t^2 + 22t - 1)\gamma^6 - (129t^2 + 184t - 27)\gamma^5 + (1169t^2 - 432t - 9)\gamma^4 +$
$(5735t^2 + 348t - 3)\gamma^3 - 2t(1202t - 23)\gamma^2 - 6t(31t + 18)\gamma + 900t^2$

and

$b(\gamma) = \ 15t\gamma^4 + (45t+8)\gamma^3 - (85t+4)\gamma^2 + (19t-4)\gamma + 30t.$

Appendix B : Equivalence Classes of Treatments

The denominator of $(h_2 - h_{84})(x_2)$ is positive for all $\gamma \in (0,1)$ and $t \geq 6$.
Rewrite $a(\gamma) = a_6\gamma^6 + \ldots + a_1\gamma + a_0$. We get $a(\gamma) > 0$ for $\gamma \in (0,1)$ and $t \geq 6$ because $a_5\gamma^5 + a_4\gamma^4 > (a_5 + a_4)\gamma^4 = 2(520t^2 - 308t + 9)\gamma^2 > 0$ and $a_1\gamma + a_0 > (a_1 + a_0)\gamma = 6(119t - 18)t\gamma > 0$.
Abbreviate $b(\gamma) = B_{03}(\gamma) + 15t\gamma^4$. Since $15t\gamma^4 > 0$ for all $\gamma > 0$ and $t \geq 6$, the function $b(\gamma)$ is positive iff $B_{03}(\gamma)$ is positive on the interval $(0,1)$. Thus, differentiating $B_{03}(\gamma)$ twice, reveals that $B_{03}(\gamma)$ is concave in the interval $(0, \gamma_{inf})$, and $B_{03}(\gamma)$ is convex in interval $(\gamma_{inf}, 1)$, whereas $\gamma_{inf} = \frac{85t+4}{3(45t+8)}$. There is just one stationary point of $B_{03}(\gamma)$ located in $(0,1)$ at
$\gamma_{st} = \frac{85t+4-2\sqrt{1165t^2+191t+28}}{3(45t+8)} < \gamma_{inf}$. Hence, $B_{03}(\gamma_{st})$ is a maximum of $B_{03}(\gamma)$ and the local minimum of $B_{03}(\gamma)$ is either range boundary 0 or 1, which are both positive, because $B_{03}(\gamma \searrow 0) = 30t$ and $B_{03}(\gamma \nearrow 1) = 9t$.
It follows that $B_{03}(\gamma)$ is positive, and, therefore, $b(\gamma)$ is positive for all $\gamma \in (0,1)$ and $t \geq 6$.
Since both functions $a(\gamma)$ and $b(\gamma)$ are positive, $(h_2 - h_{84})(x_2) > 0$ is equivalent to

$$a^2(\gamma) - 4b^2(\gamma)RT_{x_2} > 0$$
$$\Leftrightarrow 3\gamma^2[(23t-1)\gamma + 7t + 1]^2 \cdot g(\gamma) > 0$$
$$\Leftrightarrow g(\gamma) := g_8\gamma^8 + \ldots + g_1\gamma + g_0 > 0,$$

in which

$g(\gamma) = \quad 75(5t^2 + 2t + 1)\gamma^8 + 40(36t^2 + 4t - 3)\gamma^7 - 2(1171t^2 - 384t - 9)\gamma^6 -$
$\quad\quad 4(1835t^2 + 399t - 6)\gamma^5 + (23935t^2 + 850t + 3)\gamma^4 - 4t(1561t - 87)\gamma^3 -$
$\quad\quad 8t(521t + 69)\gamma^2 + 8t(517t - 16)\gamma + 1008t^2.$

The coefficients g_8 and g_7 are positive for $t \geq 6$. Observe that $g_6\gamma^6 + g_5\gamma^5 + 9935t^2\gamma^4 > (g_6 + g_5 + 9935t^2)\gamma^4 = (253t^2 - 828t + 42)\gamma^4 > 0$ for $t \geq 6$, such that $g(\gamma)$ is positive iff $G_{04}(\gamma) := (g_4 - 9935t^2)\gamma^4 + g_3\gamma^3 + g_2\gamma^2 + g_1\gamma + g_0 > 0$ for all $\gamma \in (0, 0.35]$, $\gamma \in (0.35, 1)$ and $t \geq 6$.

d1) There is one point of inflection for $G_{04}(\gamma)$ in the interval $(0,1)$ at

$$\gamma_{inf} = \frac{3t(1561t - 87) + \sqrt{3t(36486163t^3 + 4820558t^2 + 263559t + 828)}}{3(14000t^2 + 850t + 3)}.$$

$G_{04}(\gamma)$ is concave for $\gamma \in (0, \gamma_{inf})$. Thus, the local minimum of $G_{04}(\gamma)$ is either $G_{04}(\gamma \searrow 0) = g_0$, which is positive, or $G_{04}(\gamma_{inf})$. We have $\gamma_{inf} > 0.35$ and $G_{04}(0.35) \gg 85t(t - 1) > 0$. Therefore, $G_{04}(\gamma)$ is positive for all $\gamma \in (0, 0.35]$ and all $t \geq 6$.

d2) In order to analyze the performance of $G_{04}(\gamma)$ for $\gamma \in (0.35, 1)$, apply the transformation $\gamma = 0.35 + 0.65\gamma'$, in which $\gamma' \in (0, 1)$. As a result, $G_{04}(\gamma)$ converts to

$$G_{04}(\gamma') = [28561(14000t^2 + 850t + 3)\gamma'^4 + 8788(66780t^2 + 7690t + 21)\gamma'^3 - 338(86840t^2 - 87630t - 441)\gamma'^2 + 52(2649460t^2 - 481470t + 1029)\gamma' + 21(14380160t^2 - 645670t + 343)]/160000.$$

Rewrite $G_{04}(\gamma') = [g'_4 \gamma'^4 + \ldots + g'_1 \gamma' + g'_0]/160000$. Since $\gamma'^2 < \gamma'$, we get $g'_2 \gamma'^2 + g'_1 \gamma' > (g'_2 + g'_1)\gamma' \gg 78\gamma' > 0$ for all $\gamma' \in (0, 1)$ and $t \geq 6$. All other coefficients g_0, g_1, g_3 and g_4 are positive. Thus, it is concluded that $G_{04}(\gamma')$ is positive for $\gamma' \in (0, 1)$. Hence, $G_{04}(\gamma)$ is positive for all $\gamma \in (0.35, 1)$ and $t \geq 6$.

Making use of the results d1) and d2), it follows that $G_{04}(\gamma)$ is positive for all $\gamma \in (0, 1)$ and, therefore, the function $g(\gamma)$ is positive for all $\gamma \in (0, 1)$ and $t \geq 6$.
We arrive at $h_2(x_2) > h_{84}(x_2)$ for all $\gamma \in (0, 1)$ and $t \geq 6$.

e) Further, analyze

$$(h_2 - h_{38})(x) = \frac{3(2\gamma^3 + 10\gamma^2 + 15\gamma + 3)}{(2\gamma + 3)(\gamma^2 + 5\gamma + 3)} - \frac{3\gamma^3 + 19\gamma^2 + 17\gamma + 3}{(2\gamma + 3)(\gamma^2 + 5\gamma + 3)} x$$
$$+ \frac{-3(3t - 1)\gamma^4 - 3(19t - 1)\gamma^3 - (101t + 3)\gamma^2 - (25t + 3)\gamma + 12t}{2t(1 - \gamma)(2\gamma + 3)(\gamma^2 + 5\gamma + 3)} x^2.$$

Substitution of $x = x_2$ provides

$$(h_2 - h_{38})(x_2) = \frac{(1 - \alpha)a(\gamma) - 2b(\gamma)\sqrt{RT_{x_2}}}{\gamma^2(\gamma^2 + 5\gamma + 3)[(23t - 1)\gamma + 7t + 1]^2},$$

in which

$$a(\gamma) = -15(21t^2 + 22t - 1)\gamma^5 + 2(182t^2 - 85t - 6)\gamma^4 + (2055t^2 + 166t - 3)\gamma^3 - 2t(2381t + 4)\gamma^2 - 6t(361t + 33)\gamma + 900t^2$$

and

$$b(\gamma) = 15t\gamma^4 + 4(11t + 2)\gamma^3 - (174t + 1)\gamma^2 - 7(11t + 1)\gamma + 30t.$$

The denominator of $(h_2 - h_{38})(x_2)$ is positive for all $\gamma \in (0, 1)$ and $t \geq 6$.
The following subitems e1) and e2) show that the functions $a(\gamma)$ and $b(\gamma)$ have both one root in interval $(0, 1)$.

e1) Define $a(\gamma) = A_{02}(\gamma) + A_{35}(\gamma)$ and rewrite $A_{02}(\gamma) = a_0 + a_1\gamma + a_2\gamma^2$ and $A_{35}(\gamma) = a_3\gamma^3 + a_4\gamma^4 + a_5\gamma^5$.
The first derivative of $A_{35}(\gamma)$ is $A'_{35}(\gamma) = \gamma^3(6046t^2 - 1832t + 18) > 0$ for $\gamma \in (0, 1)$ and $t \geq 6$. Thus, $A_{35}(\gamma)$ is monotonous and increasing in γ in the interval $(0, 1)$.

Appendix B : Equivalence Classes of Treatments

Additionally, $A_{35}(\gamma)$ is positive because $A_{35}(\gamma) > (a_5 + a_4 + a_3)\gamma^4 = 2t(1052t - 167)\gamma^4 > 0$ for all $\gamma \in (0,1)$ and $t \geq 6$.

Use the fact that a_1 and a_2 are negative and observe that the first and second derivatives of $A_{02}(\gamma)$ are negative for $\gamma \in (0,1)$ and $t \geq 6$. Thus, function $A_{02}(\gamma)$ is monotonous and decreasing in γ in the interval $(0,1)$. The function values of the range boundaries, $A_{02}(\gamma \searrow 0) = a_0 > 0$ and $A_{02}(\gamma \nearrow 1) \ll -2t < 0$, have different algebraic signs. As a consequence, $A_{02}(\gamma)$ has exactly one root in the interval $(0,1)$. The sum of two monotonous functions with different algebraic signs in its slopes and one function offering one root in the interval $(0,1)$, can only have two roots at most iff the function values of the range boundaries have the same algebraic sign. Iff not, the sum of the two functions has exactly one root. The range boundaries of $a(\gamma)$ are 0 and 1 with function values $a(\gamma \searrow 0) = a_0 > 0$ and $a(\gamma \nearrow 1) = -36t(109g+15) < 0$ for $t \geq 6$. Hence, $a(\gamma)$ has one root at γ_a in interval $(0,1)$. In more detail, we get $\gamma_a \in (0.27, 0.28)$.

e2) Similar to e1), rewrite $b(\gamma) = B_{02}(\gamma) + B_{34}(\gamma)$ and define $B_{02}(\gamma) = b_0 + b_1\gamma + b_2\gamma^2$ and $B_{34}(\gamma) = b_3\gamma^3 + b_4\gamma^4$.

The coefficients b_3 and b_4 are positive. Thus, $B_{34}(\gamma)$ is positive, monotonous and increasing in γ for $\gamma \in (0,1)$.

Equivalent to $A_{02}(\gamma)$, function $B_{02}(\gamma)$ is monotonous and decreasing in γ because the coefficients b_1 and b_2 are negative. The function values of $B_{02}(\gamma \searrow 0) = b_0 > 0$ and $B_{02}(\gamma \nearrow 1) = -(221t+8) < 0$ have different algebraic signs. Thus, $B_{02}(\gamma)$ has one root in interval $(0,1)$.

Again, the function $b(\gamma)$ is the sum of a monotonous, increasing positive function and a monotonous, decreasing function with one root. The function values of $b(\gamma \searrow 0) = b_0 > 0$ and $b(\gamma \nearrow 1) = -162t < 0$ have different algebraic signs. Hence, $b(\gamma)$ has exactly one root at γ_b in interval $(0,1)$. More precisely, $\gamma_b \in (0.25, 0.26)$.

In order to determine whether $(h_2 - h_{38})(x_2)$ is positive, the intercept points $\gamma_a > \gamma_b$ conceal three cases:

1. Case: $\gamma \in (0, \gamma_b)$, which implies that $b(\gamma)$ and $a(\gamma)$ are positive. The expression $(h_2 - h_{38})(x_2) > 0$ is equivalent to

$$(1-\gamma)^2 a^2(\gamma) - 4b^2(\gamma) RT_{x_2} > 0$$
$$\Leftrightarrow 3\gamma^2(1-\gamma)[(23t-1)\gamma + 7t + 1]^2 \cdot g(\gamma) > 0$$
$$\Leftrightarrow g(\gamma) := g_7\gamma^7 + \ldots + g_1\gamma + g_0 > 0.$$

2. Case: $\gamma \in (\gamma_a, 1)$, which implies that $b(\gamma)$ and $a(\gamma)$ are negative. The expression $(h_2 - h_{38})(x_2) > 0$ is equivalent to

$$4b^2(\gamma) RT_{x_2} - (1-\gamma)^2 a^2(\gamma) > 0$$
$$\Leftrightarrow -3\gamma^2(1-\gamma)[(23t-1)\gamma + 7t + 1]^2 \cdot g(\gamma) > 0$$
$$\Leftrightarrow -g(\gamma) := -(g_7 \gamma^7 + \ldots + g_1 \gamma + g_0) > 0.$$

3. Case: $\gamma \in [\gamma_b, \gamma_a]$, which implies that $b(\gamma) < 0$ and $a(\gamma) > 0$. The instant conclusion is: $(h_2 - h_{38})(x_2) > 0$.

Function $g(\gamma)$ of Cases 1 and 2 is given as

$$\begin{aligned}g(\gamma) = \;& -75(5t^2 + 2t + 1)\gamma^7 - 5(311t^2 + 14t - 9)\gamma^6 + (5295t^2 + 170t + 27)\gamma^5 + \\ & (20503t^2 + 1754t + 3)\gamma^4 - 4t(8768t - 91)\gamma^3 - 4t(7095t + 419)\gamma^2 + \\ & 8t(358t - 49)\gamma + 1728t^2.\end{aligned}$$

In order to evaluate the performance of $g(\gamma)$, define functions $G_{03}(\gamma) := g_3 \gamma^3 + g_2 \gamma^2 + g_1 \gamma + g_0$ and $G_{47}(\gamma) := g(\gamma) - G_{03}(\gamma)$.

Since both coefficients g_2 and g_3 are negative, the second derivative $G''_{03}(\gamma)$ is negative for $\gamma \in (0,1)$ and $t \geq 6$. The function values of $G_{03}(\gamma)$ at $\gamma \searrow 0$ and $\gamma \nearrow 1$ are $G_{03}(0) = g_0 > 0$ and $G_{03}(1) = -12t(4905t + 142) < 0$. Thus, $G_{03}(\gamma)$ has one root in interval $(0, 1)$, because $G_{03}(\gamma)$ is concave and the function values of the range boundaries have different algebraic signs. Recall, that a concave function, like $G_{03}(\gamma)$, which is restricted to $(0, 1)$ and a positive function value at the left range boundary must be monotonous and decreasing in γ for all $G_{03}(\gamma) < 0$, regardless of an potential maximum of the function in the interval $(0, 1)$.

Use the fact that $g_7 \gamma^7 + g_6 \gamma^6 + g_5 \gamma^5 > (g_7 + g_6 + g_5)\gamma^5 = (3365t^2 - 50t - 3)\gamma > 0$ and it follows that the function $G_{47}(\gamma)$ is positive for all $\gamma \in (0, 1)$ and $t \geq 6$. A similar approach leads to the conclusion that $G_{47}(\gamma)$ is increasing in γ because $G'_{47}(\gamma) > (7g_7 + 6g_6 + 5g_5)\gamma^4 + 4g_4 \gamma^3 = 20(726t^2 - 31t - 6)\gamma^4 + 4(20503t^2 + 1754t + 3)\gamma^3 > 0$ for all $\gamma \in (0, 1)$ and $t \geq 6$.

To summarize, the function $g(\gamma)$ is the sum of a positive, monotonous increasing function $G_{47}(\gamma)$ and a concave function $G_{03}(\gamma)$ with one root in the interval $(0, 1)$. $G_{03}(\gamma)$ is also monotonous decreasing for all $G_{03}(\gamma) < 0$. Adding a positive increasing function to $G_{03}(\gamma) < 0$, $g(\gamma)$ can have two roots at most in the interval $(0, 1)$ iff the function values of $g(\gamma)$ of the range boundaries have the same sign. Iff not, $g(\gamma)$ has exactly one root. The function values of the range boundaries are $g(\gamma \searrow 0) = g_0 > 0$ and $g(\gamma \nearrow 1) = -34992t^2 < 0$. Thus, $g(\gamma)$ has exactly one root. It proves that $g(0.26) > (37t - 200)t$ is

Appendix B : Equivalence Classes of Treatments

positive and $g(0.27) < -t(142t-211)$ is negative for $t \geq 6$. It can be concluded that $g(\gamma)$ is positive for $\gamma \in (0, 0.26)$ and $g(\gamma)$ is negative for $\gamma \in (0.27, 1)$. Hence, assuming Case 1, $g(\gamma)$ is positive, assuming Case 2, $-g(\gamma)$ is positive. The performance of the function $g(\gamma)$ implies that $(h_2 - h_{38})(x_2)$ is positive. Thus, $h_2(x_2) > h_{38}(x_2)$ for all $\gamma \in (0,1)$ and $t \geq 6$.

f) A further relation to analyze is

$$(h_2 - h_{55})(x) = \frac{(4\gamma^2 + 11\gamma + 5)}{(\gamma+1)(2\gamma+3)} - \frac{(9\gamma^2 + 13\gamma + 6)}{3(\gamma+1)(2\gamma+3)}x$$
$$+ \frac{-3(7t-1)\gamma^3 - 2(32t+1)\gamma^2 - (23t+5)\gamma + 18t}{6t(1-\gamma)(\gamma+1)(2\gamma+3)}x^2.$$

Substitution of $x = x_2$ provides

$$(h_2 - h_{38})(x_2) = \frac{a(\gamma) - b(\gamma)\sqrt{RT_{x_2}}}{3\gamma^2(\gamma+1)[(23t-1)\gamma + 7t + 1]^2},$$

in which

$a(\gamma) = \;\;3(61t^2 + 89t - 4)\gamma^5 - 2(673t^2 + 76t - 12)\gamma^4 + 2(2597t^2 - 277t - 6)\gamma^3 + 2t(389t + 383)\gamma^2 - 3t(1153t + 109)\gamma + 1350t^2$

and

$b(\gamma) = \;\;3(31t-1)\gamma^3 - 2(74t-13)\gamma^2 - (155t + 23)\gamma + 90t$.

The denominator of $(h_2 - h_{55})(x_2)$ is positive for all $\gamma \in (0,1)$ and $t \geq 6$. Once again, the algebraic signs of functions $a(\gamma)$ and $b(\gamma)$ need to be determined in order to evaluate whether $(h_2 - h_{38})(x_2)$ is positive or not.

f1) Rewrite $a(\gamma) = a_5\gamma^5 + \ldots + a_1\gamma + a_0$ and define $A_{03}(\gamma) = a_0 + a_1\gamma + a_2\gamma^2 + (a_3 - 2194t^2)\gamma^3$. The function $a(\gamma)$ is positive iff $A_{03}(\gamma)$ is positive because $a_5 > 0$ and $a_4\gamma^4 + 2194t^2\gamma^3 > (a_4 + 2194t^2)\gamma^3 = 8(106t^2 - 19t + 3)\gamma^3 > 0$ for all $\gamma \in (0,1)$ and $t \geq 6$.

The coefficients a_2 and $(a_3 - 2194t^2)$ are positive for $t \geq 6$. Thus, the second derivative $A_{03}''(\gamma)$ is positive for $\gamma \in (0,1)$ and the function $A_{03}(\gamma)$ is convex in the entire interval $(0,1)$. There is one stationary point of $A_{03}(\gamma)$ in interval $(0,1)$, located at

$$\gamma_{st} = \frac{\sqrt{2t(15868142t^3 - 806981t^2 - 40621t - 5886)} - 2t(389t + 383)}{6(1500t^2 - 277t - 6)}.$$

Since $A_{03}(\gamma)$ is a convex function, $A_{03}(\gamma_{st})$ is a minimum. Some simple equation transformations confirm that

$$A_{03}(\gamma_{st}) = \frac{Vt^2 - t\sqrt{2t(15868142t^3 - 806981t^2 - 40621t - 5886)^3}}{27(1500t^2 - 277t - 6)^2}.$$

is positive for $t \geq 6$. The substitute V of $A_{03}(\gamma_{st})$ is given as $V = 100412893976t^4 - 13346937327t^3 + 823674048t^2 - 44746465t - 5450814$. The conclusion is: $A_{03}(\gamma)$ is positive for all $\gamma \in (0,1)$ and, therefore, $a(\gamma)$ is positive for all $\gamma \in (0,1)$ and $t \geq 6$.

f2) The curvature of $b(\gamma)$ switches from positive to negative, i.e., the function $b(\gamma)$ is concave for all $\gamma < \gamma_{inf} := \frac{2(74t-13)}{9(31t-1)}$ and convex for all $\gamma > \gamma_{inf}$. We have $\gamma_{inf} > 0.5$, such that local minima of $b(\gamma)$ in the interval $(0, 0.5)$ are located in the range boundaries itself. Since $b(\gamma)$ is concave in the interval $(0, 0.5)$, $b(\gamma \searrow 0) = 90t > 0$ and $b(0.5) = -(103t + 43)/8 < 0$, the function $b(\gamma)$ has root in interval $(0, 0.5)$. Furthermore, $b(\gamma)$ must be decreasing for all $\gamma \in [0.5, \gamma_{inf}]$ and is convex for all $\gamma \in (\gamma_{inf}, 1)$. As $b(\gamma \nearrow 1) = -120t < 0$, the entire function must be negative for all $\gamma \in [0.5, 1)$. Thus, $b(\gamma)$ has exactly one root in $(0, 1)$ at γ_b. Narrowing the root, we arrive at $\gamma_b \in (0.44, 0.45)$, such that $b(\gamma)$ is positive iff $\gamma \in (0, \gamma_b)$, and negative otherwise.

Knowing that $a(\gamma)$ is positive for $\gamma \in (0, 1)$, expression $a(\gamma) - b(\gamma)\sqrt{RT_{x_2}}$ is positive for all $b(\gamma) < 0$. Thus, assume $b(\gamma) > 0$. The inequality $a(\gamma) - b(\gamma)\sqrt{RT_{x_2}} > 0$ is equivalent to

$$a^2(\gamma) - b^2(\gamma)RT_{x_2} > 0$$
$$\Leftrightarrow 6\gamma^2[(23t-1)\gamma + 7t + 1]^2 \cdot g(\gamma) > 0$$
$$\Leftrightarrow g(\gamma) := g_6\gamma^6 + \ldots + g_1\gamma + g_0 > 0,$$

in which

$g(\gamma) = 24(63t^2 + 1)\gamma^6 - (6601t^2 - 455t + 48)\gamma^5 + 2(2525t^2 - 610t + 12)\gamma^4 + 6t(1575t + 151)\gamma^3 - 4t(1898t - 97)\gamma^2 - t(2593t + 529)\gamma + 2124t^2$.

The intention is to prove that $g(\gamma)$ is positive for all $\gamma \in (0, 0.5)$, since $(0, \gamma_b)$ is a subset of $(0, 0.5)$, cf. property f2). Rewrite $g(\gamma) = \gamma^4 G_{46}(\gamma) + G_{03}(\gamma)$ to get $G_{46}(\gamma) := g_4 + g_5\gamma + g_6\gamma^2$ and $G_{03}(\gamma) := g_0 + g_1\gamma + g_2\gamma^2 + g_3\gamma^3$.

The function $G_{46}(\gamma)$ is monotonous and decreasing for all $\gamma \in (0, 1)$ because its first derivative, $G'_{46}(\gamma) = g_5 + 2g_6\gamma < g_5 + g_6 = -7t(511t - 65)$ is negative for $t \geq 6$. The local minimum of $G_{46}(\gamma)$ in interval $(0, 1)$ is $G_{46}(0.5) = (4255t^2 - 1985t + 12)/2 > 0$ for $t \geq 6$. Thus, $G_{46}(\gamma)$ is positive for all $\gamma \in (0, 0.5)$.

Analyze now $G_{03}(\gamma)$. Its curvature is negative and positive as well, i.e., $G_{03}(\gamma)$ is concave iff $\gamma < \gamma_{inf2} := \frac{2(1898t-97)}{9(1575t+151)}$ and convex iff $\gamma > \gamma_{inf2}$. There is one stationary point of

Appendix B : Equivalence Classes of Treatments

$G_{03}(\gamma)$ in the interval $(0,1)$, located at

$$\gamma_{st} = \frac{\sqrt{2(65575007t^2 + 8076766t + 794183)} + 4(1898t - 97)}{18(1575t + 151)}.$$

Use the fact that $\gamma_{inf2} < \gamma_{st}$ and $\gamma_{st} > 0.5$ to get that $G_{03}(\gamma_{st})$ is a minimum of $G_{03}(\gamma)$ and $G_{03}(\gamma)$ is decreasing for $\gamma \in (0, 0.5)$. Hence, $G_{03}(\gamma)$ is positive for all $\gamma \in (0, 0.5)$ because $G_{03}(0.5) = t(443t^{-}217)/4 > 0$ for $t \geq 6$.
The conclusion is: $g(\gamma) = \gamma^4 G_{46}(\gamma) + G_{03}(\gamma)$ is positive for all $\gamma \in (0, 0.5)$ and, therefore, for all $\gamma \in (0, \gamma_b)$ as well.
We arrive at $h_2(x_2) > h_{55}(x_2)$ for all $\gamma \in (0,1)$ and $t \geq 6$.

g) The last expression to examine is

$$(h_2 - h_{114})(x) = \frac{6(5\gamma^2 + 12\gamma + 3)}{(2\gamma + 3)(5\gamma + 3)} - \frac{15\gamma^2 + 13\gamma + 6}{(2\gamma + 3)(5\gamma + 3)}x$$
$$+ \frac{-15(3t-1)\gamma^3 - 3(37t-11)\gamma^2 - 4(2t-3)\gamma + 24t}{2t(1-\gamma)(2\gamma+3)(5\gamma+3)}x^2.$$

Substitution of $x = x_2$ provides

$$(h_2 - h_{38})(x_2) = \frac{a(\gamma) - 2b(\gamma)\sqrt{RT_{x_2}}}{\gamma^2(5\gamma+3)[(23t-1)\gamma + 7t + 1]^2},$$

in which

$a(\gamma) = \ 75(21t^2 + 22t - 1)\gamma^5 + 2(421t^2 + 1192t - 6)\gamma^4 + (11403t^2 - 886t + 69)\gamma^3 - 2(1819t^2 - 649t - 9)\gamma^2 - 6t(497t - 159)\gamma + 1800t^2$

and

$b(\gamma) = \ 75t\gamma^3 - 13(16t - 5)\gamma^2 - (49t - 31)\gamma + 60t.$

The denominator of $(h_2 - h_{114})(x_2)$ is positive for all $\gamma \in (0,1)$ and $t \geq 6$.
In order to show that the function $a(\gamma)$ is positive for all $\gamma \in (0,1)$, rewrite $a(\gamma) = a_5\gamma^5 + \ldots + a_1\gamma + a_0$. The coefficients a_5 and a_4 are positive for $t \geq 6$. Thus, $a(\gamma) > 0$ is equivalent to $A_{03}(\gamma) := a_0 + a_1\gamma + a_2\gamma^2 + a_3\gamma^3 > 0$ for all $\gamma \in (0,1)$.
Derive $A_{03}(\gamma)$ twice and find $A''_{03}(\gamma) < 0$ iff $\gamma < \gamma_{inf} := \frac{2(1819t^2 - 649t - 9)}{3(11403t^2 - 886t + 69)}$, and $A''_{03}(\gamma) > 0$ iff $\gamma > \gamma_{inf}$. There is just one stationary point of $A_{03}(\gamma)$ in interval $(0,1)$ at

$$\gamma_{st} = \frac{\sqrt{2W} + 2(1819t^2 - 649t - 9)}{3(11403t^2 - 886t + 69)},$$

in which $W = 57623141t^4 - 25002895t^3 + 2353421t^2 - 75375t + 162$. Since $\gamma_{inf} < \gamma_{st}$ and $A_{03}(\gamma)$ is convex for all $\gamma > \gamma_{st}$, the value $A_{03}(\gamma_{st})$ is a local minima of the function

$A_{03}(\gamma)$. Observe that

$$A_{03}(\gamma_{st}) = \frac{4[V - \sqrt{2W^3}]}{27(11403t^2 - 886t + 69)^2},$$

whereas $V = 1277432961431t^6 - 9755571192t^5 - 26892004035t^4 + 2413576147t^3 - 48111462t^2 - 2035125t + 2916$. Some simple calculus confirms $A_{03}(\gamma_{st}) > 0$ for $t \geq 6$. Hence, $A_{03}(\gamma)$ is positive and, therefore, $a(\gamma)$ is positive for all $\gamma \in (0,1)$ and all $t \geq 6$. The function $b(\gamma)$ is monotonous and decreasing for all $\gamma \in (0,1)$. Use the fact that $\gamma^2 < \gamma$ to get $b'(\gamma) < 225t\gamma - 26(16t-5)\gamma - (49t-31)\gamma < -(240t-161)\gamma < 0$ for all $\gamma \in (0,1)$ and $t \geq 6$. Thus, the slope of $b(\gamma)$ is negative in interval $(0,1)$. The function values of the range boundaries $\gamma \searrow 0$ and $\gamma \nearrow 1$ have different algebraic signs. Hence, $b(\gamma)$ has one root in interval $(0,1)$. Since $b(0.5) = (254 - 57t)/8 < 0$ for $t \geq 6$, we conclude that $b(\gamma)$ is negative for all $\gamma \in [0.5, 1)$.

Iff $b(\gamma)$ is negative, the numerator $a(\gamma) - 2b(\gamma)\sqrt{RT_{x_2}}$ is positive because $a(\gamma)$ is positive for all $\gamma \in (0,1)$. The immediate conclusion is that $(h_2 - h_{55})(x_2)$ is positive.

For all γ with $b(\gamma) > 0$, the numerator $a(\gamma) - 2b(\gamma)\sqrt{RT_{x_2}} \stackrel{!}{>} 0$ is equivalent to

$$a^2(\gamma) - 4b^2(\gamma) RT_{x_2} > 0$$
$$\Leftrightarrow 3\gamma^2[(23t-1)\gamma + 7t + 1]^2 \cdot g(\gamma) > 0$$
$$\Leftrightarrow g(\gamma) := g_6\gamma^6 + \ldots + g_1\gamma + g_0 > 0,$$

in which

$g(\gamma) = 1875(5t^2 + 2t + 1)\gamma^6 - 50(1025t^2 - 304t - 87)\gamma^5 + (118735t^2 - 15414t + 3423)\gamma^4 - 4(1867t^2 - 2713t - 261)\gamma^3 - 4(14440t^2 - 4463t - 27)\gamma^2 + 16t(716t + 235)\gamma + 6912t^2.$

Restrict γ to the interval $(0, 0.5)$. This will include all function values of $b(\gamma)$ which are positive. Hence, iff $g(\gamma)$ is positive for $\gamma \in (0, 0.5)$, the function $g(\gamma)$ is positive for all $b(\gamma) > 0$. Use the fact that $g_6 > 0$ and $g_5\gamma^5 + 51250t^2\gamma^4 > (g_5 + 51250t^2)\gamma^4 \geq 0$ for $t \geq 6$ and $\gamma \in (0,1)$. It follows that $g(\gamma)$ is positive iff $G_{04}(\gamma) := (g_4 - 51250t^2)\gamma^4 + g_3\gamma^3 + g_2\gamma^2 + g_1\gamma + g_0 > 0$ for $\gamma \in (0, 0.5)$.

For this purpose, differentiate $G_{04}(\gamma)$ twice and observe that $G_{04}(\gamma)$ has one point of inflection in interval $(0, 0.5)$ which is located at

$$\gamma_{inf} = \frac{\sqrt{Z} + 1867t^2 - 2713t - 261}{3(22495t^2 - 5138t + 1141)},$$

whereas $Z = 653141289t^4 - 359306152t^3 + 83984933t^2 - 8490928t + 6507$. $G_{04}(\gamma)$ is concave for all $\gamma \in (0, \gamma_{inf})$ and convex for all $\gamma \in (\gamma_{inf}, 0.5)$. The alteration from concavity to

Appendix B : Equivalence Classes of Treatments

convexity implies that $G'_{04}(\gamma_{inf})$ is nonpositive. Iff $G'_{04}(0.5) \leq 0$, the function $G_{04}(\gamma)$ must be decreasing in $(\gamma_{inf}, 0.5)$ as there is no further point of inflection located in this particular interval. Therefore, calculate the function value of the slope $G'_{04}(0.5) = -(36325t^2 - 44088t - 5205)/2 < 0$ for $t \geq 6$. Hence, the local minimum of $G_{04}(\gamma)$ in interval $(0, 0.5)$ is either range boundary $\gamma \searrow 0$ or $\gamma \nearrow 0.5$. The function values of $G_{04}(\gamma)$ at 0 and 0.5 are $6912t^2 > 0$ and $(23749t^2 + 107778t + 5943)/16 > 0$, respectively, for all $t \geq 6$.

It follows that $g(\gamma)$ is positive for all $\gamma \in (0, 0.5)$ because $G_{04}(\gamma)$ is positive for all $\gamma \in (0, 0.5)$ and $g(\gamma) - G_{04}(\gamma)$ is positive for all $\gamma \in (0, 1)$. Add the fact that $a(\gamma) - 2b(\gamma)\sqrt{RT_{x_2}}$ is positive for $\gamma \in [0.5, 1)$ and the final conclusion of this property g) is: $h_2(x_2) > h_{114}(x_2)$ for all $\gamma \in (0, 1)$ and $t \geq 6$.

□

As Proof of Lemma 12. ...

b) Next to analyze is

$$(h_{28} - h_1)(x_{min}) = \frac{g(\gamma)}{3(5t-1)^2(1-\gamma)(\gamma+1)(\gamma+3)},$$

in which

$$g(\gamma) = 301t^2 - 119t + 12)\gamma^3 + 3(125t^2 - 53t + 6)\gamma^2 - (610t^2 - 245t + 24)\gamma - 3(6t^2 - 11t + 2).$$

The denominator of $(h_{28} - h_1)(x_{min})$ is positive for all $\gamma \in (0, 1)$ and $t \geq 6$. The numerator $g(\gamma) := g_3\gamma^3 + g_2\gamma^2 + g_1\gamma + g_0$ performs as follows. Differentiate $g(\gamma)$ twice and it gives $g''(\gamma) > 0$ for $t \geq 6$ and $\gamma \in (0, 1)$ since g_3 and g_2 are both positive. The positive coefficients imply that $g(\gamma)$ is convex and can have up to two roots in the interval $(0, 1)$. The function $g(\gamma)$ is continuous and it follows that $g(\gamma)$ has exactly one root at $\gamma_{root} \in (0, 1)$ because $g(\gamma \searrow 0) = g_0$ is negative and $g(\gamma \nearrow 1) = 48t^2$ is positive for $t \geq 6$. Thus,

$$(h_{28} - h_1)(x_{min}) \begin{cases} < 0 & \Leftrightarrow \gamma < \gamma_{root} \\ \geq 0 & \Leftrightarrow \gamma \geq \gamma_{root} \end{cases}.$$

B.2 Sequence Length $p = 6$

Further, calculate

$$g(\gamma_{\alpha 2}) = -\frac{V + Z\sqrt{W}}{8(613t^2 - 239t + 24)^3},$$

in which

$V = $ $255180863548t^8 - 410488247348t^7 + 287443352916t^6 - 114504471500t^5 + 28390200837t^4 - 4486763522t^3 + 441322859t^2 - 24691062t + 601128,$

$W = $ $339889t^4 - 317254t^3 + 108061t^2 - 16140t + 900$

and

$Z = $ $380807860t^6 - 454733736t^5 + 226099812t^4 - 60016064t^3 + 8983427t^2 - 720025t + 24180.$

Remodel $g(\gamma_{\alpha 2})$ properly to get $g(\gamma_{\alpha 2}) < 0$. However, the negative output implies that $\gamma_{\alpha 2} < \gamma_{root}$.

Hence, $h_1(x_{min}) > h_{28}(x_{min})$ for all $\gamma \in (\gamma_{\alpha 1}, \gamma_{\alpha 2})$.

c) The next expression to examine is

$$(h_{170} - h_1)(x_{min})$$
$$= \frac{5(5t-1)(29t-6)\gamma^2 - (671t^2 - 255t + 24)\gamma - 2(12t^2 - 17t + 3)}{6(5t-1)^2(1-\gamma)(\gamma+1)}.$$

The denominator of $(h_{170} - h_1)(x_{min})$ is positive for all $\gamma \in (0, 1)$ and $t \geq 6$. Abbreviate the numerator of $(h_{170} - h_1)(x_{min})$ as function $g(\gamma)$. There is just one root of $g(\gamma)$ in interval $(0, 1)$. This root is located at

$$\gamma_{root} = \frac{\sqrt{519841t^4 - 469130t^3 + 157633t^2 - 23400t + 1296} + 671t^2 - 255t + 24}{10(5t-1)(29t-6)},$$

such that $g(\gamma)$ is negative for all $\gamma \in (0, \gamma_{root})$. Some simple equivalent equation transformations confirm that $\gamma_{\alpha 2} < \gamma_{root}$.

Hence, $h_1(x_{min}) > h_{170}(x_{min})$ for all $\gamma \in (\gamma_{\alpha 1}, \gamma_{\alpha 2})$.

d) Calculate

$$(h_{84} - h_1)(x_{min}) = \frac{g(\gamma)}{6(5t-1)^2(1-\gamma)(\gamma^2 + 5\gamma + 3)},$$

in which

$g(\gamma) = $ $5(5t-1)(29t-6)\gamma^3 + (1793t^2 - 727t + 78)\gamma^2 - 2(1070t^2 - 427t + 42)\gamma - 12(21t^2 - 14t + 2).$

The denominator of $(h_{84} - h_1)(x_{min})$ is positive for all $\gamma \in (0, 1)$ and $t \geq 6$. Abbreviate $g(\gamma) := g_3\gamma^3 + g_2\gamma^2 + g_1\gamma + g_0$. The second derivative of $g(\gamma)$ is positive for all $\gamma \in (0, 1)$

Appendix B : Equivalence Classes of Treatments

because g_3 and g_2 are positive for $t \geq 6$. Thus, $g(\gamma)$ is convex in the entire interval $(0,1)$. The convexity implies that the local maximum of $g(\gamma)$ is located in one of the range boundaries of interest, which is either $\gamma_{\alpha 1}$ or $\gamma_{\alpha 2}$. Since $0 < \gamma_{\alpha 1} < \gamma_{\alpha 2} < \gamma_{\beta(t)} < 0.96$, $(\gamma_{\alpha 1}, \gamma_{\alpha 2})$ is a subset of $[0, 0.96]$. Observe that $g(\gamma \searrow 0) = -12(21t^2 - 14t + 2) < 0$ and $g(0.96) = -(29980507t^2 - 135917873t + 14857122)/2391250 < 0$. Thus, $g(\gamma)$ is negative in the entire interval $[0, 0.96]$ which implies that $g(\gamma)$ is negative for all $\gamma \in (\gamma_{\alpha 1}, \gamma_{\alpha 2})$. To summarize, $h_1(x_{min}) > h_{84}(x_{min})$ for all $\gamma \in (\gamma_{\alpha 1}, \gamma_{\alpha 2})$.

e) Next,
$$(h_{38} - h_1)(x_{min}) = \frac{g(\gamma)}{6(5t-1)^2(1-\gamma)(\gamma^2 + 5\gamma + 3)},$$
whereas
$$g(\gamma) = 5(5t-1)(29t-6)\gamma^3 + (1799t^2 - 727t + 78)\gamma^2 - 2(965t^2 - 409t + 42)\gamma - 12(4t-1)(9t-2).$$

The denominator of $(h_{38} - h_1)(x_{min})$ is positive for all $\gamma \in (0,1)$ and $t \geq 6$. Abbreviate the numerator $g(\gamma)$ as $g_3\gamma^3 + g_2\gamma^2 + g_1\gamma + g_0$. The second derivative of $g(\gamma)$ is positive for all $\gamma \in (0,1)$ because g_3 and g_2 are positive for $t \geq 6$. Thus, $g(\gamma)$ is convex in the entire interval $(0,1)$. The convexity implies that the local maximum of $g(\gamma)$ is located in one of the range boundaries, $\gamma_{\alpha 1}$ or $\gamma_{\alpha 2}$. Since $0 < \gamma_{\alpha 1} < \gamma_{\alpha 2} < 0.952$, the interval $(\gamma_{\alpha 1}, \gamma_{\alpha 2}) \subset [0, 0.952]$. Consequently, $g(0) = g_0 < 0$ and $g(0.952) = -6(128685971t^2 - 666365608t + 71054112)/57671875 < 0$. Thus, $g(\gamma)$ is negative in the entire interval $[0, 0.952]$ which implies that $g(\gamma)$ is negative for all $\gamma \in (\gamma_{\alpha 1}, \gamma_{\alpha 2})$. It follows that $h_1(x_{min}) > h_{38}(x_{min})$ for all $\gamma \in (\gamma_{\alpha 1}, \gamma_{\alpha 2})$.

f) Another expression to analyze is
$$(h_{55} - h_1)(x_{min})$$
$$= \frac{2(143t^2 - 58t + 6)\gamma^2 - (7t-2)(19t-3)\gamma - 3(46t^2 - 19t + 2)}{3(5t-1)^2(1-\gamma)(\gamma+1)}.$$

The denominator of $(h_{55} - h_1)(x_{min})$ is positive for all $\gamma \in (0,1)$ and $t \geq 6$. Abbreviate the numerator of $(h_{55} - h_1)(x_{min})$ as function $g(\gamma)$. There is just one root of $g(\gamma)$ in the interval $(0,1)$. This root is located at
$$\gamma_{root} = \frac{\sqrt{175561t^4 - 144934t^3 + 45013t^2 - 6228t + 324} + (7t-2)(19t-3)}{4(143t^2 - 58t + 6)},$$
such that $g(\gamma)$ is negative for all $\gamma \in (0, \gamma_{root})$. Some simple equivalent equation transformations confirm $\gamma_{\alpha 2} < \gamma_{root}$.
Hence, $h_1(x_{min}) > h_{55}(x_{min})$ for all $\gamma \in (\gamma_{\alpha 1}, \gamma_{\alpha 2})$.

g) Now observe the expression

$$(h_{114} - h_1)(x_{min})$$
$$= \frac{25(5t-1)(29t-6)\gamma^2 - (2629t^2 - 1061t + 108)\gamma - 6(144t^2 - 63t + 7)}{6(5t-1)^2(1-\gamma)(5\gamma+3)}.$$

The denominator of $(h_{114} - h_1)(x_{min})$ is positive for all $\gamma \in (0,1)$ and $t \geq 6$. Abbreviate the numerator of $(h_{114} - h_1)(x_{min})$ as function $g(\gamma)$. There is just one root of $g(\gamma)$ in the interval $(0,1)$. This root is located at

$$\gamma_{root} = \frac{\left[\sqrt{W} + 2629t^2 - 1061t + 108\right]}{50(5t-1)(29t-6)},$$

in which $W = 19439641t^4 - 16157338t^3 + 5051185t^2 - 703776t + 36864$. The function $g(\gamma)$ is negative for all $\gamma \in (0, \gamma_{root})$. Some simple equivalent equation transformations confirm $\gamma_{\alpha 2} < \gamma_{root}$.

Hence, $h_1(x_{min}) > h_{114}(x_{min})$ for all $\gamma \in (\gamma_{\alpha 1}, \gamma_{\alpha 2})$.

□

Appendix B : Equivalence Classes of Treatments

Appendix C : Some more Lemmata and Transformations

C.1 Lemmata

Lemma 13. The parameter $a_{n_j(l)}$ is increasing as $n_j(l)$ increases, $j = 1, \ldots, t$.

Proof. The assumption $a_1 = 1 < 1/(1-\gamma^2) = a_2$ holds because $\gamma \in (0,1)$. ✓

$$a_{n_j(l)} < a_{n_j(l)+1}$$
$$\Leftrightarrow \frac{(n_j(l)-2)\gamma+1}{[(n_j(l)-1)\gamma+1](1-\gamma)} < \frac{(n_j(l)-1)\gamma+1}{[n_j(l)\gamma+1](1-\gamma)}$$
$$\Leftrightarrow (n_j(l)\gamma - 2\gamma + 1)(n_j(l)\gamma + 1) < (n_j(l)\gamma - \gamma + 1)^2$$
$$\Leftrightarrow 2n_j(l)\gamma - 2\gamma < \gamma^2 + 2(n_j(l)-1)\gamma$$
$$\Leftrightarrow 0 < \gamma^2 \ ✓$$

The statement follows from the assumption and the definition of $a_{n_j(l)}$. □

Lemma 14. Assuming $n_j(l) \geq 2$, $b_{n_j(l)}$ is increasing as $n_j(l)$ increases and is maximal for $n_j(l) = 1$, $j = 1, \ldots, t$.

Proof. $b_{n_j(l)} < b_{n_j(l)+1} < 0 = b_1$ holds, because the enumerator of $b_{n_j(l)}$ is a negative constant and the denominator is an increasing function of $n_j(l)$ $\forall n_j(l) > 1$. □

Lemma 15. The column sum $crs_{n_j(l)}$ of S_{du}^{-1} of treatment j is decreasing as $n_j(l)$ increases for all $j = 1, \ldots, t_l$. On the contrary, $n_j(l) crs_{n_j(l)}$ is increasing as $n_j(l)$ increases.

Proof. 1) We have

$$crs_{n_j(l)} = a_{n_j(l)} + [n_j(l) - 1]b_{n_j(l)} = \frac{1}{[n_j(l)-1]\gamma + 1},$$

Appendix C : Some more Lemmata and Transformations

which is an decreasing function of $n_j(l)$.

2)

$$n_j(l) crs_{n_j(l)} \overset{!}{<} (n_j(l) + 1) crs_{n_j(l)+1}$$
$$\Leftrightarrow \frac{n_j(l)}{(n_j(l) - 1)\gamma + 1} < \frac{n_j(l) + 1}{n_j(l)\gamma + 1}$$
$$\Leftrightarrow 0 < 1 - \gamma,$$

which is true, as $\gamma \in (0,1)$.

Lemma 15 follows from 1) and 2). □

Lemma 16. R_u is increasing as the number of treatments t_l in sequence u increases.

Proof. R_u is the sum of all matrix entries of S_{du}^{-1}, therefore, it can be written as the sum of all row sums of S_{du}^{-1}:

$$R_u = \sum_{r=1}^{p} crs_{n_{j(r)}(l)}.$$

The sum is maximal iff every summand is maximal. Thus, R_u is maximal iff every summand equals 1, i.e., $crs_{n_j(l)} = 1$ for every treatment j. Since $crs_{n_j(l)}$ decreases as $n_j(l)$ increases, R_u increases as the $n_j(l)$ decrease. R_u is maximal iff all $n_j(l) = 1$, for every $j = 1, \ldots, t_l$. A decreasing $n_j(l)$ means that another $n_{j*}(l)$ increases under the condition of $n_{j*}(l) \leq n_j(l)$ and $j^* \in \{1, \ldots, t\}$. Thus, iff t_l increases, there must be some $n_j(l)$ which decreases, $j \in \{1, \ldots, t_l\}$, and any n_{j*} increases to a value unequal to 0, $j \in \{1, \ldots, t\} \setminus \{1, \ldots, t_l\}$.

Lemma 16 follows. □

Note, rearranging $\sum_{r=1}^{p} crs_{n_{j(r)}(l)}$, then R_u can also be written as

$$R_u = \sum_{i=1}^{t} n_j(l) crs_{n_j(l)}. \tag{C.1}$$

Lemma 17. The intersection of h_1 and h_2 at x_2 of (3.13) is less than 1 for all $t \geq p \geq 3$ and $\gamma \in (0,1)$.

Proof.

$$x_2 < 1$$
$$\Leftrightarrow \sqrt{RT_{x_2}} <$$
$$[(2p^2 - 4p - 2)t - 2]\gamma^2 + [(2p+2)t + 2]\gamma + tp(p-1)(1-\gamma) + 2t\gamma(1-\gamma)$$

Square both sides of the inequality and divide them by 4γ, gives

$$\begin{aligned}K(\gamma) :=& k_3\gamma^3 + k_2\gamma^2 + k_1\gamma + k_0\\ =& [-(2p^4 - 8p^3 + 3p^2 + 10p + 3)t^2 + (3p^2 - 6p - 4)t - 1]\gamma^3\\ &+ [(2p^4 - 11p^3 + 7p^2 + 18p + 6)t^2 - (5p^2 - 12p - 8)t + 2]\gamma^2\\ &+ [-(p^4 - 6p^3 + 6p^2 + 10p + 3)t^2 + (3p^2 - 8p - 4)t - 1]\gamma\\ &+ pt[-(p^2 - p - 2)t - p + 2]\\ &\stackrel{!}{<} 0\end{aligned}$$

a) $k_0 < 0$ because $-p + 2 < 0$ and $-(p^2 - p - 2) < 0$ for all $p \geq 3$. ✓

b) $k_1 + 1 < 0$
$\stackrel{it}{\Leftrightarrow} (*) := p^3 t \underbrace{(6-p)}_{\leq 0, p \geq 6} - 3p^2(2t-1) - 10pt - 3t - 8p - 4 < 0$
Case $p = 3$: $(*) = -6t - 1 < 0$ for all $t \geq 3$
Case $p = 4$: $(*) = -11t + 12 < 0$ for all $t \geq 3$
Case $p = 5$: $(*) = -78t + 31 < 0$ for all $t \geq 3$
$\Rightarrow k_1 + 1 < 0 \Rightarrow k_1 < 0 \; \forall t \geq p \geq 3$. ✓

c) $k_2 - 2 > 0$ for all $p \geq 5$
$\Leftrightarrow (*) := p^3 t^2 \underbrace{(2p-11)}_{>0, p \geq 6} + p^2 t(7t-5) + 6pt(3t+2) + 6t^2 + 8t > 0$
Case $p = 3$: $(*) = -12t^2 - t < 0$ for all $t \geq 3$
Case $p = 4$: $(*) = -2t^2 - 24t + 2 < 0$ for all $t \geq 3$
Case $p = 5$: $(*) = 146t^2 - 57t + 2 < 0$ for all $t \geq 3$
$\Rightarrow k_2 - 2 > 0 \; \forall p \geq 5 \Rightarrow k_2 > 0 \; \forall p \geq 5$. ✓

d) $k_3 + 1 < 0$ for all $p \geq 4$
$\Leftrightarrow (*) := 2p^3 t^2 \underbrace{(4-p)}_{\leq 0, p \geq 4} - 3p^2 t(p-1) - 2pt(3+5t) - t(3t-4) < 0$
Case $p = 3$: $(*) = t(21t - 68) < 0$ iff $t = 3$, or
$(*) = t(21t - 68) > 0$ for all $t > 3$
$\Rightarrow k_3 + 1 < 0$ for all $p > 3$, $\Rightarrow k_3 < 0 \; \forall t \geq p > 3$. ✓

Rewrite $K(\gamma) = A(\gamma) \cdot \gamma + k_0$, $A(\gamma)$ is continuous expandable on $\gamma \in [0, 1]$, i.e., $A(0) = k_1 < 0$ and $A(1) = -pt[(p^3 - 3p^2 + 2p + 2)t - p + 2] < 0$ because $[(p^3 - 3p^2 + 2p + 2)t - p + 2] \stackrel{t \geq 3}{>} p^3 - 3p^2 + p + 4 \stackrel{p \geq 3}{>} p + 4 > 0$.

Appendix C : Some more Lemmata and Transformations

1) $A''(\gamma) = 2k_3 \stackrel{d)}{>} 0$ iff $p = t = 3$. Hence, $A(\gamma)$ is convex (parabola) and local maxima of $A(\gamma)$ are given in the range boundaries $\gamma \searrow 0$ and $\gamma \nearrow 1$. $A(0)$ and $A(1)$ are negative. Since $A(\gamma)$ is continuous, it is negative for all $\gamma \in (0,1)$. ✓

2) $A''(\gamma) = 2k_3 \stackrel{d)}{<} 0$ for all $t \geq p > 3$, and it causes $A(\gamma)$ to be concave. Local maxima of $A(\gamma)$ are given in its turning points, or, if those turning points are not in the interval $(0,1)$, in the range boundaries of $A(\gamma)$ as well.
$A'(\gamma) = 2k_3\gamma + k_2 \stackrel{!}{=} 0 \Leftrightarrow \gamma = \gamma_{max} = -k_2/(2k_3)$.
Iff $k_2 < 0$ (see item c)), $\gamma_{max} < 0$ and, thus, out of range $(0,1)$. Hence, $A(\gamma)$ is monotonous and the local maxima are given by $A(0)$ and $A(1)$ which are both negative.
Else, for $k_2 > 0$, $\gamma_{max} > 0$ and $A(\gamma_{max})$ is the maximum of function $A(\gamma)$. It is

$$\begin{aligned}A(\gamma_{max}) =& k_1 - k_2^2/(4k_3) \\ =& -pt[(p-3)(4p^6 - 24p^5 + 31p^4 + 31p^3 - 36p^2 - 36p - 8)t^3 \\ & - 2(p-2)(8p^4 - 33p^3 + 13p^2 + 44p + 14)t^2 \\ & + 5(p-2)(3p^2 - 6p - 4)t - 4(p-2)] \\ & / \left(4[(p^2 - 2p - 1)t - 1][(2p^2 - 4p - 3)t - 1]\right)\end{aligned}$$

The denominator is positive for all $t \geq p \geq 3$. Next, rewrite the enumerator as $-pt \cdot G(t)$ with $G(t) = g_3 t^3 + g_2 t^2 + g_1 t + g_0$.
It is $g_0 < 0$ and $g_1 > 0$. Therefore, $g_0 + g_1 t > g_0 + g_1 = 3(p-2)(5p^2 - 10p - 8) > 0$ for all $p \geq 3$.
Further, $g_2 < 0$ and $g_3 > 0$. Thus, $g_2 + g_3 t > g_2 + 4g_3 = 2(p^2 - 2p - 2)(8p^5 - 56p^4 + 102p^3 + 17p^2 - 99p - 38) > 0$ for all $t \geq 4$.
Because of the factor $-pt$, of the enumerator of $A(\gamma_{max})$, the maximum of $A(\gamma)$ is negative for all $t \geq p \geq 4$. Since $A(\gamma)$ is concave, it is negative for all $\gamma \in (0,1)$. ✓

The function $A(\gamma)$ is negative (see items 1) and 2)), the coefficient k_0 is negative (see a)) and $\gamma \in (0,1)$. Hence, $K(\gamma) = A(\gamma) \cdot \gamma + k_0$ is negative. It follows, $x_2 < 1$. □

Lemma 18. The intersection of h_1 and h_k at x_k is positive and less than 1 for all $t \geq p \geq 5$ and $\gamma \in (0.3, 1)$.

Proof.A) Assume p is odd and $\gamma \in (0.3, 1)$.
The denominator of x_k of equation 3.14 is increasing in γ because its first derivative is positive for all γ in the domain $(0.3, 1)$ and all $t \geq p \geq 5$. The outcome of the specified denominator

at $\gamma = 0.3$ is $[(90p^3 - 553p^2 + 2136p - 273)t + 273p - 273]/500 > 0$. Thus, the denominator of x_k is positive for all $\gamma \in (0.3, 1)$ and all $t \geq p \geq 5$. It follows:
A1)

$$x_k \stackrel{!}{>} 0 \Leftrightarrow$$

$$4(p-1)pt(1-\gamma)[\gamma^2 + (p-1)\gamma + 1] \cdot [((p^2 - 2p + 1)t - p + 1)\gamma^3 +$$
$$+ (p^3 - 2p^2 + p)t\gamma^2 + ((2p^2 - 3p - 1)t + p - 1)\gamma + (3p - p^2)t] > 0 \Leftrightarrow$$
$$g(\gamma) := [((p^2 - 2p + 1)t - p + 1)\gamma^3 + (p^3 - 2p^2 + p)t\gamma^2 +$$
$$+ ((2p^2 - 3p - 1)t + p - 1)\gamma + (3p - p^2)t] > 0$$

Transform γ by $\gamma = 0.3 + 0.7\gamma'$, in which $\gamma' \in (0, 1)$ and $\gamma \in (0.3, 1)$. Thus, $g(\gamma)$ alters to

$g(\gamma') = \frac{1}{1000}[343(p-1)((p-1)t-1)\gamma'^3 + 49(p-1)((10p^2-p-9)t-9)\gamma'^2 + 7((60p^3 + 107p^2 - 294p - 73)t + 73(p-1))\gamma' + (90p^3 - 553p^2 + 2136p - 273)t + 273(p-1)]$,

which is positive for all $t \geq p \geq 5$ and all $\gamma' \in (0, 1)$. ✓

A2)

$$x_k \stackrel{!}{<} 1 \Leftrightarrow$$
$$-4 \cdot a(\gamma) \cdot b(\gamma) < 0 \Leftrightarrow$$
$$a(\gamma) > 0 \wedge b(\gamma) > 0 \text{ or } a(\gamma) < 0 \wedge b(\gamma) < 0$$

in which

$a(\gamma) = [(p^2-2p+1)t-p+1]\gamma^3 + (p^3-2p^2+p)t\gamma^2 + [(2p^2-3p-1)t+p-1]\gamma + (3p-p^2)t$

and

$b(\gamma) = [(2p^2 - 5p + 3)t - p + 1]\gamma^3 + (2p^3 - 8p^2 + 8p)t\gamma^2 + [(-2p^3 + 11p^2 - 12p - 3)t + p - 1]\gamma + (p^3 - 5p^2 + 8p)t$.

Again, apply the transformation $\gamma = 0.3 + 0.7\gamma'$ with $\gamma' \in (0, 1)$ and $\gamma \in (0.3, 1)$. The functions $a(\gamma)$ and $b(\gamma)$ alter to $a(\gamma')$, being identical to $g(\gamma')$ of passage A1), and

$b(\gamma') = [343(p-1)((2p-3)t-1)\gamma'^3 + 49((20p^3 - 62p^2 + 35p + 27)t - 9(p-1))\gamma'^2 - 7((80p^3 - 674p^2 + 855p + 219)t - 73(p-1))\gamma' + (580p^3 - 2366p^2 + 4985p - 819)t + 273p - 273]/1000$,

respectively. The function $a(\gamma)$ is positive for all $\gamma \in (0.3, 1)$ and $t \geq p \geq 5$ because all coefficients of γ'^i, $0 \leq i \leq 3$, are positive. Rewrite $b(\gamma')$ as $b_3\gamma'^3 + \ldots + b_0$. We get $b_1\gamma' + b_0 > (b_1 + b_0)\gamma' > 0$ for all $t \geq p \geq 5$. The coefficients b_2 and b_3 are positive for all $t \geq p \geq 5$ as well. Thus, $b(\gamma)$ is positive for all $\gamma \in (0.3, 1)$ and all $t \geq p \geq 5$. ✓

Lemma 18 for x_k of equation (3.14) follows from combining properties A1) and A2).

Appendix C : Some more Lemmata and Transformations

B) Assume p is even and $\gamma \in (0.3, 1)$.
The denominator of x_k of equation 3.15 is increasing in γ as the first derivative is positive for all γ in the domain $(0.3, 1)$ and all $t \geq p \geq 5$. The function value of the specified denominator at $\gamma = 0.3$ is $[(90p^3 - 616p^2 + 2545p - 546)t + 273p - 546]/500 > 0$. Thus, the denominator of x_k is positive for all $\gamma \in (0.3, 1)$ and all $t \geq p \geq 5$. It follows:
B1)

$$x_k \stackrel{!}{>} 0 \Leftrightarrow$$

$$4(p-2)pt(1-\gamma)[2\gamma^2 + (p-1)\gamma + 1] \cdot [((2p^2 - 5p + 2)t - p + 2)\gamma^3 +$$
$$+ (p^3 - 3p^2 + 2p)t\gamma^2 + ((2p^2 - 5p - 2)t + p - 2)\gamma + (4p - p^2)t] > 0 \Leftrightarrow$$
$$g(\gamma) := [((2p^2 - 5p + 2)t - p + 2)\gamma^3 + (p^3 - 3p^2 + 2p)t\gamma^2 +$$
$$+ ((2p^2 - 5p - 2)t + p - 2)\gamma + (4p - p^2)t] > 0$$

Transform γ by $\gamma = 0.3 + 0.7\gamma'$, in which $\gamma' \in (0, 1)$ and $\gamma \in (0.3, 1)$. Thus, $g(\gamma)$ alters to
$g(\gamma') = \frac{1}{1000}[343(p-2)((2p-1)t-1)\gamma'^3 + 49(p-2)((10p^2+8p-9)t-9)\gamma'^2 + 7((60p^3 + 74p^2 - 515p - 146)t + 73(p-2))\gamma' + (90p^3 - 616p^2 + 2545p - 546)t + 273p - 546]$,
which is positive for all $t \geq p \geq 5$ and all $\gamma' \in (0, 1)$. ✓
B2)

$$x_k \stackrel{!}{<} 1 \Leftrightarrow$$
$$-4 \cdot a(\gamma) \cdot b(\gamma) < 0 \Leftrightarrow$$
$$a(\gamma) > 0 \wedge b(\gamma) > 0 \text{ or } a(\gamma) < 0 \wedge b(\gamma) < 0$$

in which
$a(\gamma) = \quad [(2p^2 - 5p + 2)t - p + 2]\gamma^3 + (p^3 - 3p^2 + 2p)t\gamma^2 + [(2p^2 - 5p - 2)t + p - 2]\gamma + (4p - p^2)t$

and
$b(\gamma) = \quad [(4p^2 - 11p + 6)t - p + 2]\gamma^3 + (2p^3 - 12p^2 + 18p)t\gamma^2 + [(-2p^3 + 14p^2 - 23p - 6)t + p - 2]\gamma + (p^3 - 6p^2 + 12p)t$.

Again, apply the transformation $\gamma = 0.3 + 0.7\gamma'$ with $\gamma' \in (0, 1)$ and $\gamma \in (0.3, 1)$. The functions $a(\gamma)$ and $b(\gamma)$ alter to $a(\gamma')$, being identical to $g(\gamma')$ of passage B1), and
$b(\gamma') = \quad [343(p-2)((4p-3)t-1)\gamma'^3 + 49((20p^3 - 84p^2 + 81p + 54)t - 9(p-2))\gamma'^2 - 7((80p^3 - 788p^2 + 1517p + 438)t - 73(p-2))\gamma' + (580p^3 - 2772p^2 + 6423p - 1638)t + 273p - 546]/1000$,
respectively. The function $a(\gamma')$ is positive for all $\gamma \in (0.3, 1)$ and $t \geq p \geq 5$ because all

coefficients of γ'^i, $0 \leq i \leq 3$, are positive. Rewrite $b(\gamma')$ as $b_3\gamma'^3 + \ldots + b_0$. We get $b_1\gamma' + b_0 > (b_1 + b_0)\gamma' > 0$ for all $t \geq p \geq 5$. The coefficients b_2 and b_3 are positive for all $t \geq p \geq 5$ as well. Thus, $b(\gamma)$ is positive for all $\gamma \in (0.3, 1)$ and all $t \geq p \geq 5$. ✓
Lemma 18 for x_k of equation (3.15) follows from combining properties B1) and B2). □

Lemma 19. The minimum of $h_1 = h_k$ is located at x_k for all $\gamma \in (0.3, 1)$.

Proof. As pointed out in Lemma 18, the denominator of x_k is positive for all $\gamma \in (0.3, 1)$ as is the denominator of x_{-k}. Thus, it follows that the abscissa x_{-k} is negative because the enumerator of x_{-k} is negative. The enumerator of x_{-k} consists of a negative root term and a positive subtrahend for all $t \geq p \geq 5$, cf. equations (3.14) and (3.15).
The rest of the proof is equivalent to property (2) of the proof of Proposition 3, just replace x_1 by x_{-k} and x_2 by x_k, respectively. □

C.2 Term Transformations

The transformation of terms of $c_{22}(l)$: Multiplying terms 2 and 4 of equation (3.11) with tR_u is equivalent to

$$\left(\sum_j (n_j - \tilde{n}_{0j})crs_{n_j}\right)^2 - R_u \sum_j (n_j - \tilde{n}_{0j})crs_{n_j} + R_u \sum_j (n_j - \tilde{n}_{0j})\tilde{n}_{0j}b_{n_j}$$

$$= \underbrace{\left(\sum_j (n_j - \tilde{n}_{0j})crs_{n_j}\right)}_{=R_u - \sum_j \tilde{n}_{0j}crs_{n_j}} \underbrace{\left(\sum_j (n_j - \tilde{n}_{0j})crs_{n_j} - R_u\right)}_{=-\sum_j \tilde{n}_{0j}crs_{n_j}} + R_u \sum_j (n_j - \tilde{n}_{0j})\tilde{n}_{0j}b_{n_j}$$

$$= \left(\sum_j \tilde{n}_{0j}crs_{n_j}\right)^2 - R_u \sum_j \tilde{n}_{0j}crs_{n_j} + R_u \sum_j n_j\tilde{n}_{0j}b_{n_j} - R_u \sum_j \underbrace{\tilde{n}_{0j}^2}_{=\tilde{n}_{0j}} b_{n_j}$$

$$\stackrel{\text{Proof of}}{\underset{\text{Lemma 15}}{=}} \left(\sum_j \tilde{n}_{0j}crs_{n_j}\right)^2 - R_u \sum_j \tilde{n}_{0j}a_{n_j}.$$

This results into equation (3.12) of section 3.3.

Appendix C : Some more Lemmata and Transformations

Bibliography

Afsarinejad, K. & Hedayat, A. S. (2002): Repeated Measurements Designs for a Model with Self and Simple Mixed Carryover Effects. Journal of Statistical Planning and Inference, 106(1-2), 449–459.

Brockhoff, P. M. (1998): Assessor Modelling. Food Quality and Preference, 9(3), 87–89.

Cox, D. R. (1958): Planning of experiments. John Wiley & Sons, New York.

Han, L. (2007): Models with Subject by Treatment and Subject by Carryover Interactions and Use of Baseline Measurements in Crossover Trials. PhD Thesis. University of Georgia, Athens, Georgia.

Kemmler, G. W. (1990): Optimale Designs für Crossover-Versuche unter Einbeziehung der Anzahl der Behandlungsperioden. Dissertation. Universität Dortmund.

Kiefer, J. (1975): Construction and Optimality of Generalized Youden Designs. In: Srivastava, J. N., ed., A Survey of Statistical Design and Linear Models. North-Holland, Amsterdam, pages 333–353.

Kunert, J. (1991): Cross-over Designs for Two Treatments and Correlated Errors. Biometrika, 78(2), 315–324.

Kunert, J. & Martin, R. J. (2000a): On the Determination of Optimal Designs for an Interference Model. Annals of Statistics, 28(6), 1728–1742.

Kunert, J. & Martin, R. J. (2000b): Optimaltiy of type I orthogonal arrays for cross-over models with correlated errors. Journal of Statistical Planning and Inference, 87, 119–124.

Kunert, J. & Stufken, J. (2002): Optimal Crossover Designs in a Model With Self and Mixed Carryover Effects. Journal of the American Statistical Association, 97(459), 898–906.

Bibliography

Kushner, H. B. (1997): Optimal Repeated Measurements Designs: the Linear Optimality Equations. Annals of Statistics, 25(6), 2328–2344.

Lundahl, D. S. & McDaniel, M. R. (1988): The Panelist Effect - Fixed or Random? Journal of Sensory Studies, 3, 113–121.

Matthews, J. N. S. (1988): Recent Developments in Crossover Designs. International Statistical Review, 56(2), 117–127.

Næs, T. (1991): Handling Individual Differences between Assessors in Sensory Profiling. Food Quality and Preference, 2, 187–199.

Næs, T. & Langsrud, O. (1998): Fixed or Random Assessors in Sensory Profiling? Food Quality and Preference, 9(3), 145–152.

O'Mahony, M. (1986): Sensory Evaluation of Food. Marcel Dekker, New York.

Rao, C. R. & Toutenburg, H. (1995): Linear Models. Least Squares and Alternatives. Springer, New York.

Stufken, J. (1996): Optimal Crossover Designs. In: Gosh, S. & Rao, C. R., eds., Handbook of Statistics 13: Design and Analysis of Experiments. North-Holland, Amsterdam, pages 63–90.

Die VDM Verlagsservicegesellschaft sucht für wissenschaftliche Verlage abgeschlossene und herausragende

Dissertationen, Habilitationen, Diplomarbeiten, Master Theses, Magisterarbeiten usw.

für die kostenlose Publikation als Fachbuch.

Sie verfügen über eine Arbeit, die hohen inhaltlichen und formalen Ansprüchen genügt, und haben Interesse an einer honorarvergüteten Publikation?

Dann senden Sie bitte erste Informationen über sich und Ihre Arbeit per Email an *info@vdm-vsg.de*.

Sie erhalten kurzfristig unser Feedback!

VDM Verlagsservicegesellschaft mbH
Dudweiler Landstr. 99
D - 66123 Saarbrücken
www.vdm-vsg.de

Telefon +49 681 3720 174
Fax   +49 681 3720 1749

Die VDM Verlagsservicegesellschaft mbH vertritt

Printed by Books on Demand GmbH, Norderstedt / Germany